Márcio Rostirolla Adames

Spacelike Self-Similar Solutions of the Mean Curvature Flow

Márcio Rostirolla Adames

Spacelike Self-Similar Solutions of the Mean Curvature Flow

in Pseudo-Euclidean Spaces

Südwestdeutscher Verlag für Hochschulschriften

Impressum / Imprint
Bibliografische Information der Deutschen Nationalbibliothek: Die Deutsche Nationalbibliothek verzeichnet diese Publikation in der Deutschen Nationalbibliografie; detaillierte bibliografische Daten sind im Internet über http://dnb.d-nb.de abrufbar.
Alle in diesem Buch genannten Marken und Produktnamen unterliegen warenzeichen-, marken- oder patentrechtlichem Schutz bzw. sind Warenzeichen oder eingetragene Warenzeichen der jeweiligen Inhaber. Die Wiedergabe von Marken, Produktnamen, Gebrauchsnamen, Handelsnamen, Warenbezeichnungen u.s.w. in diesem Werk berechtigt auch ohne besondere Kennzeichnung nicht zu der Annahme, dass solche Namen im Sinne der Warenzeichen- und Markenschutzgesetzgebung als frei zu betrachten wären und daher von jedermann benutzt werden dürften.

Bibliographic information published by the Deutsche Nationalbibliothek: The Deutsche Nationalbibliothek lists this publication in the Deutsche Nationalbibliografie; detailed bibliographic data are available in the Internet at http://dnb.d-nb.de.
Any brand names and product names mentioned in this book are subject to trademark, brand or patent protection and are trademarks or registered trademarks of their respective holders. The use of brand names, product names, common names, trade names, product descriptions etc. even without a particular marking in this works is in no way to be construed to mean that such names may be regarded as unrestricted in respect of trademark and brand protection legislation and could thus be used by anyone.

Coverbild / Cover image: www.ingimage.com

Verlag / Publisher:
Südwestdeutscher Verlag für Hochschulschriften
ist ein Imprint der / is a trademark of
AV Akademikerverlag GmbH & Co. KG
Heinrich-Böcking-Str. 6-8, 66121 Saarbrücken, Deutschland / Germany
Email: info@svh-verlag.de

Herstellung: siehe letzte Seite /
Printed at: see last page
ISBN: 978-3-8381-3497-0

Zugl. / Approved by: Hannover, LUH, Diss., 2012

Copyright © 2012 AV Akademikerverlag GmbH & Co. KG
Alle Rechte vorbehalten. / All rights reserved. Saarbrücken 2012

Abstract

I classify spacelike self-similar shrinking solutions of the mean curvature flow in pseudo-euclidean space in arbitrary codimension, if the mean curvature vector is nonzero and the principal normal vector is parallel in the normal bundle. Moreover, I exclude the existence of such self-shrinkers in several cases. The classification is analogous to the existing classification in the euclidean case [Hui93, Smo05].

Keywords: mean curvature flow, self-similar, self-shrinker, pseudo-euclidean.

The present work was accomplished with support of CNPq, an entity of the Brazilian Government directed to the Technological and Scientific Development.

Contents

	Introduction	**5**
1	**Introduction to Pseudo-Euclidean Geometry**	**13**
	1.1 Pseudo-Euclidean Spaces and Hyperquadrics	13
	1.2 Connections on Semi-Riemannian Manifolds	20
	1.3 Structural Equations	27
2	**Hyperquadric Homotheties of the MCF**	**34**
	2.1 Hyperquadric Homotheties of the MCF	34
	2.2 Existence and Uniqueness	40
	2.2.1 Immersion in the Hyperquadric $\mathcal{H}^{n-1}(k)$ with $k > 0$. . .	40
	2.2.2 Immersion in the Hyperquadric $\mathcal{H}^{n-1}(k)$ with $k < 0$. . .	42
	2.2.3 Immersion in the Hyperquadric $\mathcal{H}^{n-1}(0)$	44
3	**Principal Normal Parallel in the Normal Bundle**	**48**
	3.1 Fundamental Equations	49
	3.2 The Compact Case	56
4	**The Non-Compact Case**	**69**
	4.1 The First Case	84
	4.2 The Second Case	95
5	**Summary**	**117**
6	**Appendix**	**119**
	6.1 Maximum Principles	119
	6.2 Foliations	120
	6.3 Geodesic Completeness	122

List of Symbols	**123**
Index	**126**

Introduction

The Mean Curvature Flow (MCF) of an immersion $F : M \to N$ of a smooth manifold M into a Riemannian manifold (N, h) is a natural way to deform this immersion into something "rounder" or "more regular". It is a smooth family of isometric immersions $F_t : M \to N$, $t \in [0, T)$[1] that satisfies

$$\frac{dF_t}{dt} = \vec{H}, \qquad F_0(x) = F(x),$$

The regularizing effect comes from $\vec{H} = \text{tr}(\nabla dF)$, which is just the Laplace-Beltrami operator $\triangle$ applied to each coordinate function of F if N has flat metric (for example the euclidean space), so that the MCF is the solution of a system of generalised heat equations.

Another motivation for this flow is that it is the negative L^2-gradient flow of the volume functional in the space of the immersions, so that it decreases the volume of the immersed manifold in the fastest possible direction.

This equation was proposed by [Mul56] to model the formation of grain boundaries in annealing metals and was also studied by [Bra78] from the viewpoint of geometric measure theory (an integral formulation of the MCF using varifolds), as written by Smoczyk in his survey on higher codimensional mean curvature flow [Smo11a]. The mean curvature flow has been studied since then by many different authors, some of them will be mentioned along this introduction. Nowadays, there are several other introductions and surveys on this subject in different contexts, for example [Whi02], [Ilm97b], [Nev11] and books [Eck04], [RS10], [Man10], [Bra78] and [Zhu02].

The mean curvature flow is the solution of a system of partial differential equations. As in [Smo11a] one calculates the linearization of the operator $\vec{H}$ on the class of smooth immersions (the trace of ∇dF) and finds out that the MCF

---

[1] We take this T as the maximal existence time for the MCF.

is a degenerate system of quasilinear parabolic partial differential equations. This system can be modified to a strictly parabolic system of equations using a tubular neighborhood or the DeTurck's trick (as in [Man10] and [Bak11b] respectively) and thence has short time existence and uniqueness guaranteed by the theory of partial differential equations if M, N are smooth and M is closed. If M is not compact, there are some results that guarantee existence and uniqueness in special cases. For example, Ecker and Huisken showed in [EH89] smooth long time existence for solutions of the MCF of entire graphs with Lipschitz initial condition and a lower bound greater than zero on the scalar product of the normal direction with the "height" direction of the graph; Ecker proved independently the long time existence for spacelike hypersurfaces in the Minkowski space $\mathbb{R}^{1,n}$ without extra conditions on the initial immersion in [Eck97]. In my dissertation, the questions of existence are not directly dealt with, but we classify a special case (self-shrinking solutions) under certain conditions in terms of minimal submanifolds.

One of the problems of the MCF is that it also produces singularities. Suppose now that the target manifold N is some Euclidean space, i. e. $\mathbb{R}^n$ with the usual scalar product. In an important work on the MCF of convex compact hypersurfaces [Hui84], Huisken showed that the supremum of the norm of the second fundamental form $\sup_M \|A\|^2$ explodes as $t \to T$ (the maximal existence time) if there is a finite time $(T < \infty)$ singularity[2]. This happens because an upper bound on the second fundamental form would imply upper bounds on all the derivatives $\nabla^{(k)} A_{ij}$ and the solution could be then extended beyond T, which is a contradiction. However, this can be done not only for hypersurfaces, but for a broader class of manifolds and in any codimension (see [Smo11a] Proposition 3.11 and Remark 3.12 or [Coo11]).

In a subsequent work [Hui90], Huisken showed, with his famous monotonicity formula, that hypersurfaces satisfying a natural[3] growth in the second fundamental form:

$$\max_M \|A\|^2 \leq \frac{C_0}{2(T-t)},$$

for some constant $C_0 > 0$, deform asymptotically near a singularity to self-

---

[2] But the main result in this paper of Huisken is that convex compact hypersurfaces are deformed to spheres after some blow up process. This result also works in other contexts, like the volume preserving MCF of hypersurfaces near a sphere [ES98], which has long time existence and converges exponentially to a sphere.

[3] This is the growth rate of some simple hypersurfaces, like spheres and cylinders.

similar solutions of the MCF after some blow up process (rescaling the surface and changing the time variable). This result depends only on the existence of some integrals with respect to the backwards heat kernel and holds, for example, if M is closed. Later Ilmanen [Ilm97a] and White [Whi94] proved that the all finite time singularities in the generalized sense of the Brakke flow [Bra78] are self-similar solutions of the MCF.

These self-similar solutions of the MCF are also called self-shrinkers to avoid confusion with other types of solutions that preserve the "form" of the surface, like self-expanders and translating solutions. They are homotheties that shrink the initial manifold and are given by the equation

$$\vec{H} = -F^\perp.$$

Because of the relation between singularities of the MCF and self-shrinkers, there is interest in classifying and giving examples of these in special cases. Abresch and Langer [AL86] gave the complete classification of the closed plane curves that shrink homothetically, they are the circles and the so called Abresch & Langer curves. These are transcendental curves given by two integers $\frac{1}{2} < \frac{m}{n} < \frac{\sqrt{2}}{2}$, where m is the rotation index and the curve closes up in n periods of the curvature. Huisken proved in [Hui93] that the self-shrinking hypersurfaces with non-negative mean curvature (compact or non-compact) are spheres, cylinders and the product of an Abresch & Langer curve with an affine space. The result of Huisken was later generalized by Smoczyk [Smo05] for higher codimensional immersions, with the assumption that the principal normal is parallel in the normal bundle and $\|\vec{H}\|_\mathbb{E} \neq 0$. A related result was found by Cao and Li [CL11] in any codimension: the self-shrinkers with $\|A\|^2 \leq 1$ are spheres, planes or cylinders. There are also Bernstein type results for self-shrinkers in higher codimension of Q. Ding and Z. Wang [DW09], who generalize works of Lu Wang [Wan09] and Huisken & Ecker [EH89]. Recently, Baker [Bak11a] proved that high codimensional self-shrinkers under certain conditions for the second fundamental tensor and mean curvature vector are spheres or cylinders.

For hypersurfaces in $\mathbb{R}^3$, there are examples of a shrinking doughnut of Angenent [Ang92] and many numerical examples of Chopp [Cho94] and Ilmanen [Ilm97b], like "punctured saddles" made of many handles crushing at the same time, which are highly unstable, depending on the surface having many symmetries. Colding and Minicozzi [CMI09] showed that the only stable singularities for smooth closed embedded surfaces in $\mathbb{R}^3$ are cylinders and spheres. For the Lagrangian MCF, Joyce, Lee and Tsui [JLT10], Anciaux [Anc06] and Wang [Wan08a] have examples. There are other results in different contexts.

The main purpose of this work is to study self-shrinkers of the MCF in higher codimension in the pseudo-euclidean case. By that we mean that the target manifold N is not a Riemannian manifold but a pseudo-euclidean space ($\mathbb{R}^n$ with some inner product $\langle \cdot, \cdot \rangle$[4] which is nondegenerate but not necessarily positive definite), so that the most interesting new case is the Minkowski space $\mathbb{R}^{1,n}$. The MCF of spacelike hypersurfaces in the Minkowski space was studied for example by Ecker [Eck97] and a generalization of it (the MCF with some forcing term to obtain a prescribed mean curvature) was considered by Ecker and Huisken [EH91]. Gerhardt [Ger08] also studies curvature flows in semi-Riemannian manifolds, specially the inverse mean curvature flow. Beyond this Bergner and Schäfer [BS11] have noted that, in order to find 2-dimensional self-similar solutions (not only the shrinking ones, but also the expanding and the translating solutions) of the mean curvature flow in the 3-dimensional Minkowski space, it is enough to find a surface of prescribed anisotropic mean curvature. Beyond this, Li & Salavessa have some results for the MCF of spacelike graphs [LS09] in product manifolds.

The first chapter of this dissertation is a short introduction on pseudo-Euclidean geometry based on O'Neill [O'N83] and some fundamental equations for immersions.

The second chapter is about the homotheties of the MCF that lie in hyperquadrics. As a test case we first consider the hyperquadrics $\mathcal{H}^{n-1}(k) := \{x \in (\mathbb{R}^n, \langle \cdot, \cdot \rangle) : \langle x, x \rangle = k\}$ for some $k \in \mathbb{R}$. They are a natural generalization of the euclidean spheres and one finds, similarly to Smoczyk's result in [Smo05] for spheres in the Euclidean space $\mathbb{E}^n$, that the homotheties (with the pullback of the metric being nondegenerate) of the MCF with initial immersion contained in a hyperquadric are exactly the minimally immersed submanifolds of the hyperquadric if $k > 0$ or $k < 0$, as Theorems 2.2.1 and 2.2.2 state. Moreover, given the initial minimal immersion, the flow can be explicitly calculated. If $k = 0$ (the light cone), a homothety with nondegenerate first fundamental form would

---
[4]We denote $\|X\|^2 := \langle X, X \rangle$.

immediately leave the light cone and thence could not[5] be a homothety starting at $t=0$ because the light cone is star shaped, as stated in Theorem ??. But, as Ecker noted in [Eck97], the upper light-cone would immediately change to a hyperquadric and the explicit solution of the MCF with the upper light-cone as initial condition in the Minkowski space $(\mathbb{R}^{1,n})$ would be the graph of the function

$$\delta(x,t) = \sqrt{\|x\|_{\mathbb{E}}^2 + 2(n-1)t},$$

which is a homothety after $t=0$.

There is a big difference between the flow of minimal surfaces of the hyperquadrics with $k>0$ and the ones with $k<0$; if $k>0$, they shrink to a point (at least the compact ones) in finite time, but if $k<0$, they expand and never produce singularities. Beyond this, they are given by different equations. The following results are for the shrinking[6] case, which are the isometric immersions $F: M \to (\mathbb{R}^n, \langle \cdot, \cdot \rangle)$ satisfying

$$\vec{H} = -F^\perp.$$

Our "domain" manifold M is always assumed to be smooth, path connected, complete and orientable.

If one considers the self shrinkers and self-expanders that are contained in the hyperquadrics as submanifolds in the pseudo-euclidean space $(\mathbb{R}^n, \langle \cdot, \cdot \rangle)$, then one observes that $\nabla^\perp \vec{H} \equiv 0$ and $\nabla^\perp \nu \equiv 0$. A natural question is whether these conditions are also sufficient to guarantee that a spacelike[7] self-shrinker lies in a hyperquadric. The condition $\nabla^\perp \vec{H} = 0$ implies this immediately if M is compact, because $\|\vec{H}\|^2$ is then constant and the maximum principle implies, with equation

$$\triangle \|F\|^2 = 2m - 2\|\vec{H}\|^2, \tag{1}$$

that $\|F\|^2$ is constant. So, in this work, we examine the self-shrinkers of the MCF with $\|\vec{H}\|^2 \neq 0$ and $\nabla^\perp \nu = 0$. The condition $\nabla^\perp \nu = 0$ is natural because it holds for any hypersurface.

The third chapter deals with fundamental equations for self-shrinkers with the principal normal parallel in the normal bundle and the compact case. The

---

[5]One could expect to find at least some stationary solutions in the light cone, like straight lines, but for such a line the metric is degenerate and thence this case is not included in Theorem 2.2.3.

[6]The expanding case satisfies $\vec{H} = F^\perp$.

[7]We use elliptic methods to obtain our results (maximum principles, that do not hold for hyperbolic equations).

fourth chapter is about the non-compact case.

Equation (1) already shows that there are no compact self-shrinkers with $\|\vec{H}\|^2 < 0$. In this dissertation, the inexistence of self-shrinkers with $\|\vec{H}\|^2 < 0$ is proven, also in the non-compact case under certain hypothesis, as stated in the following theorem:

Theorem 5.0.1 *There are no spacelike self-shrinkers $F : M \to (\mathbb{R}^n, \langle \cdot, \cdot \rangle)$ of the MCF that satisfy*

- *$F(M)$ unbounded and F is mainly negative and has bounded geometry or*

- *$\|\vec{H}\|^2 < 0$ and one of the following:*
  1. *M is compact.*
  2. *$F(M)$ is unbounded, M is stochastic complete and $\sup_M \|F\|^2 \leq \infty$.*
  3. *$F(M)$ is unbounded, F is mainly positive, has bounded geometry and the principal normal parallel in the normal bundle.*

Remark 0.0.1. The bounded geometry condition here assumed is more restrictive than the one usually assumed in the literature (see def. 4.0.14) and I use mainly positive and negative as in definition 4.0.9.

As a consequence of this, the Minkowski space does not (in all of our treated cases) have spacelike self-shrinking hypersurfaces. This could already be seen from the, already mentioned, Ecker's longtime existence result for spacelike hypersurfaces in Minkowsky space.

If $\|\vec{H}\|^2 > 0$, one finds, just as Smoczyk in [Smo05] for the Euclidean case, that if M is compact and $\dim(M) \geq 2$, the only spacelike self-shrinkers of the MCF with $\|\vec{H}\|^2(p) \neq 0$, $\forall p \in M$, and $\nabla^\perp \nu \equiv 0$ are the minimal[8] submanifolds of hyperquadrics:

Theorem 3.0.1 *Let M be a closed smooth manifold and $F : M \to (\mathbb{R}^n, \langle \cdot, \cdot \rangle)$ be a smooth immersion, which is a spacelike self-shrinker of the mean curvature*

---

[8]By minimal we mean the ones satisfying $\vec{H} = 0$. We use this name because the condition $\vec{H} = 0$ is then mnemonic, although this condition does not imply minimality of the volume functional in pseudo-euclidean spaces, so they are just critical points of the volume functional.

flow, i.e. F satisfies,
$$\vec{H} = -F^\perp. \qquad (2)$$

Besides assume $m := \dim(M) \neq 1$. Then the mean curvature vector $\vec{H}$ satisfies $\|\vec{H}\|^2(p) \neq 0$ for all $p \in M$ and the principal normal ν is parallel in the normal bundle ($\nabla^\perp \nu \equiv 0$) if, and only if, F is a minimal immersion in the hyperquadric $\mathcal{H}^{n-1}(m)$.

This dimensional restriction is in fact optimal because in dimension one there are the Abresch & Langer curves which are self-shrinkers and do not lie in spheres.

To prove this Theorem we use the maximum principle on the function

$$\frac{\|P\|^2}{\|\vec{H}\|^4}, \qquad \text{with } P := \langle A, \vec{H} \rangle,$$

where A is the second fundamental tensor of F. With this, one gets $\|P\|^2 = \|\vec{H}\|^4$ and then a nice formula for $\triangle \|\vec{H}\|^2$, which delivers, through careful consideration on the eigenvalues of P and partial integration, $\nabla \|\vec{H}\| = 0$. Then equation (1) delivers that $\|F\|^2$ is constant by the maximum principle.

Again following Smoczyk in the non-compact case, one finds that the self-shrinkers with $\|\vec{H}\|^2 > 0$ are products of affine spaces with minimal submanifolds of hyperquadrics or with solutions of the curve shortening flow[9] as stated:

Theorem 5.0.3 *Let M be a smooth manifold and $F: M \to \mathbb{R}^{q,n}$ be a mainly positive, spacelike, shrinking self-similar solution of the mean curvature flow with bounded geometry such that $F(M)$ is unbounded. Beyond that, let F satisfy the conditions: $\|\vec{H}\|^2(p) \neq 0$ for all $p \in M$ and the principal normal is parallel in the normal bundle ($\nabla^\perp \nu \equiv 0$). Then one of the two holds:*

$$F(M) = \mathcal{H}_r \times \mathbb{R}^{m-r} \quad \text{or}$$
$$F(M) = \Gamma \times \mathbb{R}^{m-1},$$

where $\mathcal{H}_r$ is an r-dimensional minimal surface of the hyperquadric $\mathcal{H}^{n-1}(r)$ (in addition $\|\vec{H}\|^2 \equiv r > 0$) and Γ is a rescaling of an Abresch & Langer curve in a

---
[9]The curve shortening flow is the MCF for plane curves.

spacelike plane. By R^{m-r} we mean an $m-r$ dimensional spacelike affine space in $\mathbb{R}^{q,n}$.

The proof of this Theorem is long and internally divided in lemmas to make its several steps easier to recognize. It was necessary to divide the proof in two cases. In both of them we split TM into two involutive distributions. Then we use the Theorem of Frobenius 6.2.4 to get foliations on M whose leaves are totally geodesic immersed in M. After this, we calculate a formula that relates the second fundamental tensor of F with these distributions. In particular the second fundamental tensor of F is zero when restricted to one of these distributions, so that the leaves of this distribution are totally geodesic in $(\mathbb{R}^{q,n}, \langle \cdot, \cdot \rangle)$ and then considering parallel transports inside these leaves, one finds that they are parallel affine subspaces of $\mathbb{R}^{q,n}$. The other distribution delivers the $\mathcal{H}_r$ and Γ parts in the last Theorem. We get this considering the second fundamental tensor and mean curvature vectors of the inclusion of the leaves related to this distribution, with some extra effort to prove that Γ lies on a plane (based on an idea of [Hui93]). In the last step we construct an explicit map from these second leaves times $\mathbb{R}^{m-r}$ onto $F(M)$.

The results called Theorems (with exception of the Theorem of Frobenius and of the Theorem of Hopf and Rinow) are new and/or generalizations.

In the appendix, there are some results used in this work that are not proven here. Before the Bibliography there is a list of symbols and after it an index.

Chapter 1

Introduction to Pseudo-Euclidean Geometry

1.1 Pseudo-Euclidean Spaces and Hyperquadrics

In this chapter, we introduce some basic definitions and properties of pseudo-euclidean spaces, most of the definitions and Theorems on this section are based on the book [O'N83] by O'Neil.

Definition 1.1.1. Let V be an n-dimensional vector space (for our pourposes $\mathbb{R}^n$). An *inner product* $\langle \cdot, \cdot \rangle$ over V is a nondegenerate symmetric bilinear form on V. This means an application $\langle \cdot, \cdot \rangle : V \times V \to \mathbb{R}$ that is

- symmetric: $\langle x, y \rangle = \langle y, x \rangle$ for all $x, y \in V$,
- bilinear: $\langle \alpha x + \beta y, w \rangle = \alpha \langle x, w \rangle + \beta \langle y, w \rangle$ for all $x, y, w \in V$ and $\alpha, \beta \in \mathbb{R}$,
- nondegenerate: $\langle x, w \rangle = 0$ for all $w \in V$ implies $x = 0$.

An inner product is said to be *positive definite* [or *negative definite*] if

$$\|x\|^2 := \langle x, x \rangle > 0 \, [\text{or} < 0] \, \forall x \in V \text{ with } x \neq 0.$$

A vector space with an inner product is called an *inner product space*.

An important number related to an inner product is the "number of directions" in which it is negative definite:

Definition 1.1.2. The *index* η of an inner product $\langle \cdot, \cdot \rangle$ over V is the maximum of the dimensions of subspaces $W \subset V$ on which $\langle \cdot, \cdot \rangle|_W$ is negative definite.

Example 1.1.3. The Euclidean space $\mathbb{E}^n$ of dimension n, which is $\mathbb{R}^n$ equipped with the usual inner product is an inner product space.

Another inner product space is the Minkowski space:

Definition 1.1.4. The *Minkowski n-space* $(\mathbb{R}^{1,n})$ is the real n-dimensional space $\mathbb{R}^n$ endowed with the inner product $\langle \cdot, \cdot \rangle$ defined through

$$\langle x, y \rangle = -x^1 y^1 + x^2 y^2 + \cdots + x^n y^n$$

for all $x, y \in \mathbb{R}^n$, $x = (x^1, \ldots, x^n), y = (y^1, \ldots, y^n)$. If we identify $T_x \mathbb{R}^{1,n} \cong \mathbb{R}^{1,n}$ for any $x \in \mathbb{R}^n$ this inner product defines a metric (that is not positive definite) on $\mathbb{R}^n$. This metric is called the *Minkowski metric*.

The Euclidean space has index 0 and the Minkowski space has index 1.

The property of being nondegenerate is fundamental to our applications and it can be characterized as follows:

Lemma 1.1.5. *A symmetric bilinear form is nondegenerate if, and only if, its matrix relative to one (hence every) basis is invertible.*

Proof. Let $\{e_1, e_2, \ldots, e_n\}$ be a basis of V and write $a_{ij} := \langle e_i, e_j \rangle$. If $v \in V$, then $\langle v, w \rangle = 0$ for all $w \in V$ if, and only if, $\langle v, e_i \rangle = 0$ for $i = 1, \ldots, n$ because of the bilinearity of $\langle \cdot, \cdot \rangle$. Then

$$\langle v, e_i \rangle = \left\langle \sum_{j=1}^n v_j e_j, e_i \right\rangle = \sum_{i=1}^n v_j a_{ij}$$

Thus, $\langle \cdot, \cdot \rangle$ is degenerated if, and only if, there are numbers $v_1, \ldots, v_n$ such that $\sum_j v_j a_{ij} = 0$ for each $i = 1, \ldots, n$, but this is exactly the linear independence of the rows of (a_{ij}) and this is equivalent to (a_{ij}) being singular. $\square$

Definition 1.1.6. Let $U \subset V$ be a vector subspace of V. The *normal subspace* $U^\perp$ is the set

$$U^\perp := \{v \in V : \langle v, u \rangle = 0 \, \forall u \in U\}$$

Given a vector subspace $U \subset V$, we would like to decompose the whole space into $V = U \oplus U^\perp$, as it is possible for subspaces in $\mathbb{E}^n$. Such decomposition is possible exactly when V is nondegenerate. The following Lemma is needed to prove this in Lemma 1.1.9.

Lemma 1.1.7. *If W is a subspace of an inner product space V, then*

1. $\dim W + \dim W^\perp = n = \dim V$

2. $(W^\perp)^\perp = W$

Proof. 1. Let $e_1, \ldots, e_n$ be a basis of V such that $e_1, \ldots, e_k$ is a basis of W and $a_{ij} := \langle e_i, e_j \rangle$ the coefficients of the matrix that represents this inner product in this basis. On the other hand $v \in W^\perp$ if, and only if, $\langle v, e_i \rangle = 0$ for $i = 1, \ldots, k$, which in coordinate terms is

$$\sum_{j=1}^n a_{ij} v_j = 0 \qquad \text{for } 1 \leq i \leq k.$$

This is a system of k-linear equations in n variables. Because of Lemma 1.1.5 the rows of the coefficient matrix are linearly independent, so that the matrix of the system has rank k. Hence the space of solutions of this system has dimension $n - k$. As the solutions of this system are exactly the vectors in $W^\perp$, it holds that $\dim W^\perp = n - k$.

2. Because of part 1 we know $\dim W^\perp + \dim(W^\perp)^\perp = n = \dim W + \dim W^\perp$ which implies that $\dim(W^\perp)^\perp = \dim W$. On the other hand $v \in (W^\perp)^\perp$ means $v \perp W^\perp$, but all elements in W are orthogonal to the elements in $W^\perp$ by definition, so that $W \subset (W^\perp)^\perp$ and, as the two subspaces have the same dimension, $W = (W^\perp)^\perp$.

$\square$

Definition 1.1.8. A subspace W of an inner product space V is called *nondegenerate* if $\langle \cdot, \cdot \rangle|_W$ is nondegenerate.

A subspace of an inner product space (with the induced bilinear form) will not always be an inner product space itself. For example if the inner product is not positive definite there will be a non-zero vector v with $\|v\|^2 = 0$ (such a vector is called a *null vector* or a *lightlike* vector) and the restriction of the inner product to the space generated by v is degenerated, thus not an inner product. The following Lemma gives a characterization of this phenomenon.

Lemma 1.1.9. *A subspace W of an inner product space V is nondegenerate if, and only if, $V = W + W^\perp$.*

Proof. From linear algebra it is true that

$$\dim(W + W^\perp) + \dim(W \cap W^\perp) = \dim W + \dim W^\perp,$$

but we know from the previous Lemma that $\dim W + \dim W^\perp = n$, so that $\dim(W + W^\perp) = n$ if, and only if, $\dim(W \cap W^\perp) = 0$. However $\dim W + \dim W^\perp = n$ if, and only if, $W + W^\perp = V$ because $W + W^\perp \subset V$. It also holds that $\dim(W \cap W^\perp) = 0$ if, and only if, given $v \in W$, $\langle v, w \rangle = 0$ for all $w \in W$ implies $v = 0$, which completes the proof. □

Remark 1.1.10. As $(W^\perp)^\perp = W$ it holds that $W + W^\perp = V$ if, and only if, $W^\perp + (W^\perp)^\perp = V$, this means that W is nondegenerate if, and only if, $W^\perp$ is nondegenerate.

It is possible to classify the vectors in an inner product space into three classes:

Definition 1.1.11. Let $v \in V$ be a vector. v is said to be

spacelike	if $\langle v, v \rangle > 0$ or $v = 0$,
null	if $\langle v, v \rangle = 0$ and $v \neq 0$,
timelike	if $\langle v, v \rangle < 0$.

We say that a vector subspace W of V is *spacelike* if all $w \in W$ are spacelike.

We would like to have an orthonormal basis for an inner product space in the sense of the Euclidean space, but this is not always possible because there can be vectors with negative length. So we consider a generalized concept of orthonormality:

Definition 1.1.12. Let V be a n-dimensional inner product space. A *unit vector* $v \in V$ is a vector such that $\|v\|^2 = \pm 1$. A set of k mutually orthogonal unit vectors is said to be *orthonormal*.

A set of n orthonormal vectors will be necessarily linearly independent and thence an orthonormal basis of V.

Lemma 1.1.13. *An inner product space $V \neq 0$ has an orthonormal basis.*

Proof. Since $\langle \cdot, \cdot \rangle$ is nondegenerate, there is a vector $v \in V$ such that $\|v\|^2 = \langle v, v \rangle \neq 0$, then
$$e_1 := \frac{v}{\sqrt{|\|v\|^2|}}$$
is a unit vector. Thus it suffices, by induction, to show that any orthonormal set $e_1, \ldots, e_k$ with $k < n := \dim(V)$ can be enlarged by one.

From the unit condition it is clear that the matrix representation of restriction of the metric $\langle \cdot, \cdot \rangle$ over the span of the set $e_1, \ldots, e_k$ is an invertible matrix. Then the Remark of Lemma 1.1.9 implies that the complement of this set is nondegenerate, so that we can choose one vector more. Which completes the proof. $\square$

We would like to be able to write a vector in this orthonormal basis.

Lemma 1.1.14. *Let $\{e_1, \ldots, e_n\}$ be an orthonormal basis for V with $\varepsilon_i := \langle e_i, e_i \rangle$. Then each $v \in V$ has a unique representation*
$$v = \sum_{i=1}^{n} \varepsilon_i \langle v, e_i \rangle e_i.$$

Proof. To prove that this is a representation of v it is enough to show that v minus this sum is orthogonal to each e_j, $j = 1, \ldots, n$, because the nondegeneracy would imply that v minus this sum is equal to zero, so
$$\left\langle v - \sum_{i=1}^{n} \varepsilon_i \langle v, e_i \rangle e_i, e_j \right\rangle = \langle v, e_j \rangle - \sum_{i=1}^{n} \varepsilon_i \langle v, e_i \rangle \langle e_i, e_j \rangle$$
$$= \langle v, e_j \rangle - \sum_{i=1}^{n} \varepsilon_i \langle v, e_i \rangle \delta_{ij} \varepsilon_i$$
$$= \langle v, e_j \rangle - \langle v, e_j \rangle = 0$$

From the linearly independence of the basis it follows that this representation is unique. $\square$

The aim of this work is to consider self-similar solutions of the mean curvature flow in a pseudo-euclidean space, i. e. a vector space $\mathbb{R}^n$ equipped with some inner product that is not positive definite. So from now on the space that we will use is $\mathbb{R}^n$ with some inner product.

In this space we can consider the set of all points with constant distance to the origin:

Definition 1.1.15. For $n \in \{2, 3, \ldots\}$ we call the set

$$\mathcal{H}^{n-1}(k) := \{x \in \mathbb{R}^n : \|x\|^2 = k\}$$

Hyperquadric of dimension $n-1$ and parameter k, $k \in \mathbb{R}$ fixed.

These hyperquadrics are just quadratic surfaces by definition. We disconsider the case $n = 1$ because these hyperquadrics would then be only two points, zero or the empty space, which are not interesting. In the Minkowski space, the hyperquadric $\mathcal{H}^{n-1}(k)$ for $k > 0$ is also called *de Sitter space* (dS_{n-1}). If $k < 0$, the hyperquadric $\mathcal{H}^{n-1}(k)$ is also called *Anti de Sitter space* (AdS_{n-1}) and the cone $\mathcal{H}^{n-1}(0)$ is called *light cone*. We use the notation $\mathcal{H}^{n-1}(k)$ because, in this work, the cases $k > 0$ and $k < 0$ are considered at once in several Propositions (despite their different geometry).

Example 1.1.16. 1) If we consider $\mathbb{R}^3$ with the canonical basis and the inner product given in this basis by the matrix A,

$$A := \begin{pmatrix} 3 & 0 & 0 \\ 0 & \frac{1}{2} & 0 \\ 0 & 0 & -1 \end{pmatrix},$$

then the piece of the hyperquadric for $k = -1$ between the planes $z = -4$ and $z = 4$ is the surface drawn in figure 1.1

2) If we consider $\mathbb{R}^3$ with the canonical basis and the inner product given in this basis by the matrix A,

$$A := \begin{pmatrix} 3 & 0 & 0 \\ 0 & \frac{1}{2} & 0 \\ 0 & 0 & 1 \end{pmatrix},$$

then the hyperquadric for $k = 1$ is the surface drawn in figure 1.2

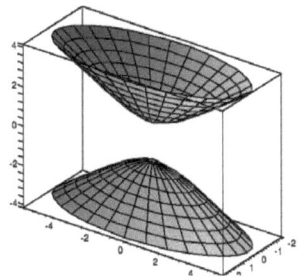

Figure 1.1: $3x^2 + \frac{1}{2}y^2 - z^2 = -1$

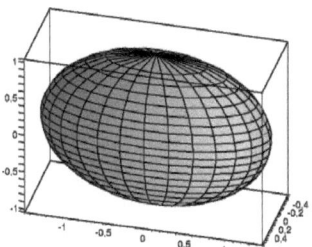

Figure 1.2: $3x^2 + \frac{1}{2}y^2 + z^2 = 1$

We want to consider the mean curvature flow of manifolds immersed in these hyperquadrics, so we first have to answer whether these objects are manifolds.

Proposition 1.1.17. *Let* $(\mathbb{R}^n, \langle \cdot, \cdot \rangle)$ *be an inner product space. If* $k \neq 0$ *then the hyperquadric* $\mathcal{H}^{n-1}(k)$ *is a smooth manifold.* $\mathcal{H}^{n-1}(0)$ *is not a smooth manifold, but* $\mathcal{H}^{n-1}(k) \setminus \{0\}$ *is.*

Proof. The inner product $\langle \cdot, \cdot \rangle$ defines a quadratic form $Q : \mathbb{R}^n \to \mathbb{R}$ through $Q(x) = \langle x, x \rangle$ for all $x \in \mathbb{R}^n$. If (a_{ij}) is the matrix that represents $\langle \cdot, \cdot \rangle$ in the canonical basis then
$$Q(x) = \sum_{ij} a_{ij} x_i x_j,$$
which is smooth. The Jacobian $J_x Q$ of Q at the point $x \in \mathbb{R}^n$ is given by
$$J_x Q = \left(2 \sum_j a_{1j} x_j, \ldots, 2 \sum_j a_{nj} x_j \right) = 2[(a_{ij})x]^T.$$

As the inner product is nondegenerate, the matrix (a_{ij}) is invertible and it follows that the only solution of $(a_{ij})x = 0$ is $x = 0$. So that each point $x \in$

$\mathbb{R}^n$, $x \neq 0$ is a regular point of Q, which implies that every $k \in \mathbb{R}$, $k \neq 0$ is a regular value of Q. Because of the Theorem of the regular value the inverse image $Q^{-1}(k)$ of a regular value k is a $n-1$ dimensional (embedded) smooth submanifold or empty. The value $0 \in \mathbb{R}$ is a regular value of Q restricted to $\mathbb{R}^n \setminus \{0\}$ and then its inverse image is a smooth submanifold (the light-cone without the origin) of $\mathbb{R}^n$. □

It is possible to generalize the notion of inner product space to differentiable manifolds.

Definition 1.1.18. Let M be a differentiable manifold and $\langle \cdot, \cdot \rangle \in \Gamma(T^*M \otimes T^*M)$ be a smooth section such that $\langle \cdot, \cdot \rangle(x)$ is an inner product for each $x \in M$. Then $\langle \cdot, \cdot \rangle$ is called a *semi-Riemannian metric* over M and the pair $(M, \langle \cdot, \cdot \rangle)$ is called a *semi-Riemannian manifold*.

Let N be a differentiable manifold, $F: N \to M$ be an immersion and $g := F^*\langle \cdot, \cdot \rangle$ the pullback of the semi-Riemannian metric over M. If g is nondegenerate, N is also a semi-Riemannian manifold. N is said to be *spacelike* if g is a Riemannian metric on N (symmetric, nondegenerate and positive-definite).

1.2 Connections on Semi-Riemannian Manifolds

In this section, we introduce some basic notions of semi-Riemannian manifolds like connections and curvatures together with some properties of them and the notation that we use.

Let $\langle \cdot, \cdot \rangle$ be an inner product on $\mathbb{R}^n$, M be a m-dimensional manifold and $F: M \to (\mathbb{R}^n, \langle \cdot, \cdot \rangle)$ be an immersion. The identification of $T_p\mathbb{R}^n$ with $\mathbb{R}^n$ induces a semi-Riemannian metric (denoted also by $\langle \cdot, \cdot \rangle$) on $\mathbb{R}^n$ and the immersion F induces a semi-Riemannian metric (also called the first fundamental form) $g := F^*\langle \cdot, \cdot \rangle$ over M, if it is nondegenerate. As we are interested in solutions of the mean curvature flow we will assume that g is nondegenerate, because the mean curvature vector is not well defined if g is degenerate. Let ∇^g be the Levi-Civita connection[1] of (M,g) and D be the Levi-Civita connection on $(\mathbb{R}^n, \langle \cdot, \cdot \rangle)$ (pointwise just the usual derivative). Then:

$$dF(\nabla^g_X Y) = (D_{dF(X)} dF(Y))^\top.$$

---
[1] The Koszul's formula (with this inner product) defines a unique metric and torsion-free connection.

The vector bundle $T\mathbb{R}^n \cong \mathbb{R}^n \times \mathbb{R}^n$ is the map $\pi : \mathbb{R}^n \times \mathbb{R}^n \to \mathbb{R}^n$ given by

$$\pi(x,V) = x \text{ for any } (x,V) \in T\mathbb{R}^n.$$

The immersion F induces the following vector bundles on M:

- The *pullback bundle* is defined with the set

$$F^*T\mathbb{R}^n := \{(p,V) \in M \times T\mathbb{R}^n \mid F(p) = \pi(V)\} \subset M \times T\mathbb{R}^n,$$

and the map $\pi' : F^*T\mathbb{R}^n \to M$, given by $\pi'(p,V) = p$.

- The *normal bundle* is defined with the set

$$TM^\perp := \{(p,V) \in M \times T\mathbb{R}^n \mid F(p) = \pi(V) \text{ and } \langle V, F_i \rangle = 0, \forall i = 1,...,m\},$$

and the map $\pi'' : TM^\perp \to M$, given by $\pi''(p,V) = p$.

We also use the following connections on several bundles:

- $\nabla^{F^*T\mathbb{R}^n}$ on the pullback bundle defined as $\nabla^{F^*T\mathbb{R}^n}_X Y := D_{dF(X)} Y$ for any $X \in \Gamma(TM)$ and $Y \in \Gamma(F^*T\mathbb{R}^n)$.

- $\nabla^\perp$ on the normal bundle defined as $\nabla^\perp_X Y = (D_{dF(X)} Y)^\perp$ for any $X \in \Gamma(TM)$ and $Y \in \Gamma(TM^\perp)$.

- ∇^* on the dual of a bundle E over M defined through $(\nabla^*_X \epsilon)(e) := X(\epsilon(e)) - \epsilon(\nabla_X e)$ for any $X \in \Gamma(TM)$, $e \in \Gamma(E)$ and $\epsilon \in \Gamma(E^*)$.

- $\nabla^{E \otimes F}$ on the product bundle $E \otimes F$ of two bundles E and F over M defined as $\nabla^{E \otimes F}_X (e \otimes f) := \nabla^E_X e \otimes f + e \otimes \nabla^F_X f$

We usually omit most of the superscript indicating where the bundle is. We just use ∇ for most of the cases and $\nabla^\perp$ if we project, on the normal bundle, the component of the tensor that lies in $TM^\perp$. For example, for $X \in \Gamma(TM)$ and $Y \otimes Z \in \Gamma(TM \otimes TM^\perp)$, it holds

$$\nabla_X(Y \otimes Z) = (\nabla_X Y) \otimes Z + Y \otimes (\nabla_X Z)$$
$$\nabla^\perp_X(Y \otimes Z) = (\nabla_X Y) \otimes Z + Y \otimes (\nabla^\perp_X Z).$$

Other important objects are the second fundamental tensor and the mean curvature vector, they are extrinsic curvatures given by an isometric immersion.

Definition 1.2.1. Let (M,g) and (N,h) be semi-Riemannian manifolds and $F: M \to N$ an immersion. We say that F is an *isometric immersion* if $g = F^*h$.

Definition 1.2.2. Let (M,g) and (N,h) be semi-Riemannian manifolds and $F: M \to N$ be an isometric immersion. The *second fundamental tensor* $A \in \Gamma(F^*TN \otimes T^*M \otimes T^*M)$ is defined as $A := \nabla dF$, with dF understood as a section in $T^*M \otimes F^*TN$. The trace of A, $\vec{H} := \mathrm{tr} A \in \Gamma(F^*TN)$, is called the *mean curvature vector*.

Remark 1.2.3. The mean curvature vector is locally written as $\vec{H} = g^{ij} A\left(\frac{\partial}{\partial x^i}, \frac{\partial}{\partial x^j}\right)$.

Remark 1.2.4. Sometimes we write A_F or $\vec{H}_F$ for the second fundamental tensor and the mean curvature vector of an isometric immersion F to make clear which immersion generates them.

Remark 1.2.5. The second fundamental tensor and the mean curvature vector are defined for isometric immersions, although we do not always make explicit mention to the metrics.

We use Latin letters for indices of tensors on M and Greek letters for indices of tensors on the target manifold N, in our case $(\mathbb{R}^n, \langle \cdot, \cdot \rangle)$ (for example the Christoffel symbols of the connection ∇^g are Γ^k_{ij} and the Christoffel symbols of the connection D are $\Gamma^\alpha_{\beta\delta} = 0$). We also use the Einstein's convention for sums. With this notation we have:

$$A_{ij} := (\nabla dF)\left(\frac{\partial}{\partial x^i}, \frac{\partial}{\partial x^j}\right) = \left(\nabla\left(\frac{\partial F^\alpha}{\partial x^k} dx^k \otimes \frac{\partial}{\partial y^\alpha}\right)\right)\left(\frac{\partial}{\partial x^i}, \frac{\partial}{\partial x^j}\right)$$
$$= \frac{\partial^2 F^\alpha}{\partial x^i \partial x^j} \frac{\partial}{\partial y^\alpha} + \frac{\partial F^\alpha}{\partial x^k}(\nabla dx^k)\left(\frac{\partial}{\partial x^i}, \frac{\partial}{\partial x^j}\right) \otimes \frac{\partial}{\partial y^\alpha}$$
$$= \frac{\partial^2 F^\alpha}{\partial x^i \partial x^j} \frac{\partial}{\partial y^\alpha} - \frac{\partial F^\alpha}{\partial x^k}\left(dx^k\left(\nabla_{\frac{\partial}{\partial x^i}} \frac{\partial}{\partial x^j}\right)\right) \otimes \frac{\partial}{\partial y^\alpha}$$
$$= \frac{\partial^2 F^\alpha}{\partial x^i \partial x^j} \frac{\partial}{\partial y^\alpha} - \frac{\partial F^\alpha}{\partial x^k} \Gamma^k_{ij} \frac{\partial}{\partial y^\alpha} = \nabla_i \nabla_j F. \tag{1.1}$$

In the case $F: M \to \mathbb{R}^n$ one can identify $\mathbb{R}^n \cong T_{F(p)} \mathbb{R}^n$ for any $p \in M$, and consider $F \in \Gamma(F^*T\mathbb{R}^n)$. This way $\nabla F \in \Gamma(T^*M \otimes F^*T\mathbb{R}^n)$ makes sense and we denote $\nabla_i \nabla_j F = \left(\nabla_{\frac{\partial}{\partial x^i}} \nabla F \right) (\frac{\partial}{\partial x^j}) = \nabla_{\frac{\partial}{\partial x^i}} \left(\nabla_{\frac{\partial}{\partial x^k}} F \otimes dx^k \right) (\frac{\partial}{\partial x^j})$. It is clear from the local representation of the second fundamental form (equation (1.1)) that it is a symmetric tensor.

Remark 1.2.6. The covariant derivative is not tensorial. But given $T \in \Gamma(T^*M^{\otimes s} \otimes TM^{\otimes r})$, ∇T is another tensor, $\nabla T \in \Gamma(T^*M^{\otimes s+1} \otimes TM^{\otimes r})$. For example, in local coordinates, if $T_i^j dx^i \otimes \frac{\partial}{\partial x^j} \in \Gamma(T^*M \otimes TM)$, we write $\nabla_l T_i^j$ for the coeficients of this new tensor $\nabla_l T_i^j dx^l \otimes dx^i \otimes \frac{\partial}{\partial x^j} \in \Gamma(T^*M \otimes T^*M \otimes TM)$, which are defined through

$$\nabla_l T_i^j \frac{\partial}{\partial x^j} := \nabla_{\frac{\partial}{\partial x^l}} \left(T_s^k dx^s \otimes \frac{\partial}{\partial x^k} \right) \left(\frac{\partial}{\partial x^i} \right). \qquad (1.2)$$

This new tensor is related to T_i^j by equation (1.2). In this work we write tensors locally but derivate in the sense of this remark, rather than deriving only the coefficients. This way to deal with tensors is called *tensor analysis* or *Ricci calculus*, one can find more details in [CLN06], pg 8.

The following product rule is then satisfied:

Lemma 1.2.7. *Let E be a fiber bundle over M, $\{e_1, \ldots, e_n\}$ a local basis to E and $\epsilon^1, \ldots, \epsilon^n$ the dual base to E^*. Besides let $A, B \in \Gamma(E^* \otimes E)$ be tensors over M, locally written $A_j^i \epsilon^j \otimes e_i$ and $B_s^r \epsilon^s \otimes e_r$, and ∇ a connection on E. Then $A_j^i B_k^j \epsilon^k \otimes e_i \in \Gamma(E^* \otimes E)$ and*

$$\nabla_l (A_j^i B_k^j) = \nabla_l A_j^i B_k^j + A_j^i \nabla_l B_k^j. \qquad (1.3)$$

Proof.

$$\nabla_l A_j^i e_i = \nabla_{\frac{\partial}{\partial x^l}} \left(A_s^i \epsilon^s \otimes e_i \right) (e_j)$$
$$= \frac{\partial A_j^i}{\partial x^l} e_i + A_s^i \nabla_{\frac{\partial}{\partial x^l}} \epsilon^s (e_j) e_i + A_s^i \epsilon^s (e_j) \nabla_{\frac{\partial}{\partial x^l}} e_i$$
$$= \frac{\partial A_j^i}{\partial x^l} e_i - A_s^i \epsilon^s \left(\nabla_{\frac{\partial}{\partial x^l}} e_j \right) e_i + A_j^i \nabla_{\frac{\partial}{\partial x^l}} e_i.$$

On the other hand, for B_k^j, we have

$$\nabla_l B_k^j e_j = \frac{\partial B_k^j}{\partial x^l} e_j - B_s^j \epsilon^s \left(\nabla_{\frac{\partial}{\partial x^l}} e_k \right) e_j + B_k^j \nabla_{\frac{\partial}{\partial x^l}} e_j,$$

so that

$$\nabla_l A^i_j B^j_k e_i = \frac{\partial A^i_j}{\partial x^l} B^j_k e_i - A^i_s B^j_k \epsilon^s \left(\nabla_{\frac{\partial}{\partial x^l}} e_j\right) e_i + A^i_j B^j_k \nabla_{\frac{\partial}{\partial x^l}} e_i$$

and

$$A^i_j \nabla_l B^j_k e_i = A^i_j \frac{\partial B^j_k}{\partial x^l} e_i - A^i_j B^j_s \epsilon^s \left(\nabla_{\frac{\partial}{\partial x^l}} e_k\right) e_i + A^i_s B^j_k \epsilon^s \left(\nabla_{\frac{\partial}{\partial x^l}} e_j\right) e_i.$$

Then finally

$$\left(\nabla_l A^i_j B^j_k + A^i_j \nabla_l B^j_k\right) e_i = \frac{\partial A^i_j}{\partial x^l} B^j_k e_i + A^i_j B^j_k \nabla_{\frac{\partial}{\partial x^l}} e_i$$
$$A^i_j \frac{\partial B^j_k}{\partial x^l} e_i - A^i_j B^j_s \epsilon^s \left(\nabla_{\frac{\partial}{\partial x^l}} e_k\right) e_i$$
$$= \nabla_l (A^i_j B^j_k) e_i.$$

$\square$

It is not hard to see that such a product rule holds for tensors with more components, with components on different bundles (as long as they can be correctly composed) and for the composition of more than two tensors.

Maximum principles play a big role by the theory of mean curvature flow. We need some Laplacian operators in this work:

Definition 1.2.8. The *Laplace Beltrami operator* (which is sometimes called just *Laplacian* in the literature) $\triangle : C^\infty(M) \to C^\infty(M)$ is defined, for $f \in C^\infty(M)$, as

$$\triangle f := \operatorname{tr}[\nabla(\nabla f)].$$

Let $F : M \to N$ be an immersion, $a, b, c \in \mathbb{N}$,

$$\Omega := \underbrace{TM \otimes \ldots \otimes TM}_{a} \otimes \underbrace{T^*M \otimes \ldots \otimes T^*M}_{b} \otimes \underbrace{F^*TN \otimes \ldots \otimes F^*TN}_{c}$$

and

$$\Psi := \underbrace{TM \otimes \ldots \otimes TM}_{a} \otimes \underbrace{T^*M \otimes \ldots \otimes T^*M}_{b} \otimes \underbrace{TM^\perp \otimes \ldots \otimes TM^\perp}_{c}$$

The *(rough) Laplacian operator* acting on tensors $\triangle : \Omega \to \Omega$ is defined, for

$X \in \Gamma(\Omega)$ as
$$\triangle X := \text{tr}[\nabla(\nabla X)].$$

The *(rough) Laplacian operator on the normal bundle* acting on tensors $\triangle^\perp : \Psi \to \Psi$ is defined, for $Y \in \Gamma(\Psi)$ as
$$\triangle^\perp Y := \text{tr}[\nabla^\perp(\nabla^\perp Y)].$$

So that in the tensor analysis notation they are written $\triangle := g^{ij}\nabla_i\nabla_j$ and $\triangle^\perp := g^{ij}\nabla_i^\perp\nabla_j^\perp$. Beyond this, using the tensor analysis notation, the second fundamental tensor $A \in \Gamma(T^*M \otimes T^*M \otimes F^*TN)$ is written $A_{ij} = \nabla_i\nabla_j F$ and it follows
$$\vec{H} = \triangle F.$$

Now we will derive some basic equations about the second fundamental tensor:

Lemma 1.2.9. *Let (M,g) be a semi-Riemannian manifold, $F:(M,g) \to (\mathbb{R}^n, \langle \cdot, \cdot \rangle)$ be an isometric immersion and A be the second fundamental tensor of F. Then $A(X,Y)(p) \in T_pM^\perp$ for all $X,Y \in \Gamma(TM)$ and $p \in M$.*

Proof. As A is a tensor, it is enough to calculate this in local coordinates at a point p. Let $\{\frac{\partial}{\partial x^1}, \dots, \frac{\partial}{\partial x^m}\}$ be a local basis to TM and $F_l := \frac{\partial F}{\partial x^l}$ be its image through dF, then

$$\langle A_{ij}, F_l \rangle = \langle \nabla_i\nabla_j F, \nabla_l F \rangle = \nabla_i g_{jl} - \langle F_j, \nabla_i\nabla_l F \rangle = -\langle F_j, A_{il} \rangle$$

which implies that
$$\langle A_{ij}, F_l \rangle + \langle A_{il}, F_j \rangle = 0,$$

it follows
$$0 = \langle A_{ij}, F_l \rangle + \langle A_{il}, F_j \rangle + \langle A_{li}, F_j \rangle + \langle A_{lj}, F_i \rangle - \langle A_{ji}, F_l \rangle - \langle A_{jl}, F_i \rangle$$
$$= 2\langle A_{il}, F_j \rangle,$$

because of the symmetry of A and, as i,j and l are arbitrary elements of the set $\{1,\dots,m\}$. $\square$

It follows immediately that:

Corollary 1.2.10. *Let (M,g) be a semi-Riemannian manifold, $F:(M,g) \to (\mathbb{R}^n, \langle \cdot, \cdot \rangle)$ be an isometric immersion and $\vec{H}$ be the mean curvature vector of F. Then $\vec{H}(p) \in T_p M^\perp$.*

Definition 1.2.11. Let (M,g) be a semi-Riemannian manifold, E be a vector bundle over M and ∇ a connection on E. The *Riemannian curvature tensor*[2] $R^\nabla \in \Gamma(E \otimes E^* \otimes T^*M \otimes T^*M)$ of M is the tensor defined for any $X, Y \in \Gamma(TM)$ and $\sigma \in E$ as

$$R^\nabla(X,Y)\sigma := \nabla_X \nabla_Y \sigma - \nabla_Y \nabla_X \sigma - \nabla_{[X,Y]}\sigma.$$

If $\langle \cdot, \cdot \rangle$ is an inner product on each fiber of E we denote, for $X, Y \in \Gamma(TM)$ and $\zeta, \sigma \in \Gamma(E)$,

$$R^\nabla(\zeta, \sigma, X, Y) := \langle \zeta, R^\nabla(X,Y)\sigma \rangle.$$

If $E = TM$, it is possible to take the trace with respect to the first and third variables of the tensor $R := R^{TM}(\cdot, \cdot, \cdot, \cdot)$ (from now on we omit the superscript in this case). This trace is called the *Ricci tensor* and written $Ric \in \Gamma(T^*M \otimes T^*M)$, so that for any $X, Y \in \Gamma(TM)$

$$Ric(X,Y) := \operatorname{tr} R(\cdot, X, \cdot, Y)$$

Taking another trace we have the *inner curvature* $s \in C^\infty(M)$

$$s := \operatorname{tr} Ric.$$

It is not hard to see that the Riemannian curvature is in fact a tensor. One can see this, for example, in the book *Riemannian Geometry* [dC92] from do Carmo for the tangent bundle or calculate it in local coordinates. We defined the Riemannian curvature tensor for vector bundles in general but we are only interested in two cases: the tangent bundle TM over M and the normal bundle $TM^\perp$.

We write the Riemannian curvature vector with respect to the tangent bun-

---

[2]O'Neill [O'N83] also uses this name, although the "Riemannian" does not have anything to do with the metric now.

dle in local coordinates:

$$R^l_{kij}\frac{\partial}{\partial x^l} = R\left(\frac{\partial}{\partial x^i},\frac{\partial}{\partial x^j}\right)\frac{\partial}{\partial x^k},$$

and

$$R_{skij} = R\left(\frac{\partial}{\partial x^s},\frac{\partial}{\partial x^k},\frac{\partial}{\partial x^i},\frac{\partial}{\partial x^j}\right) = R^l_{kij}g_{ls}.$$

We will always use the metric to lower and lift indices of tensors, for example $R^k_{l\,ij} = R_{lsij}g^{sk}$. One can see, for example in the book [dC92], from do Carmo that this tensor satisfies the following equations:

$$R_{skij} = -R_{ksij} = -R_{skji} = R_{ijsk},$$
$$R_{skij} + R_{sijk} + R_{sjki} = 0. \quad \text{(First Bianchi Identity)}$$

The Ricci tensor is written in local coordinates $R_{ij} = R_{likj}g^{lk} = R^k_{ikj}$.

1.3 Structural Equations

In order to get geometric information about $F(M)$ we need to deduce some basic relations between the tensors so far defined; these equations are called structural equations. These are the fundamental equations that give important information about the second fundamental form, the mean curvature vector and the curvatures of the manifold M like: Codazzi equation, Gauß equation and Ricci equation.

Proposition 1.3.1 (Codazzi Equation). *Let $F:(M,g) \to (\mathbb{R}^n,\langle\cdot,\cdot\rangle)$ be an immersion, $A \in \Gamma(F^*T\mathbb{R}^n \otimes T^*M \otimes T^*M)$ be the second fundamental tensor of this immersion and $X,Y,V \in \Gamma(TM)$. Then*

$$(\nabla_X A)(Y,V) - (\nabla_Y A)(X,V) = -dF(R(X,Y)V) \quad (1.4)$$

Proof. First, by definition,

$$(\nabla_X A)(Y,V) = \nabla_X(A(Y,V)) - A(\nabla_X Y,V) - A(Y,\nabla_X V),$$

but on the other hand

$$(\nabla_X \nabla_Y dF)(V) = (\nabla_X(\nabla dF(Y)))(V)$$
$$= ((\nabla_X(\nabla dF))(Y) + (\nabla dF(\nabla_X Y)))(V)$$
$$= (\nabla_X A)(Y, V) + A(\nabla_X Y, V),$$

which implies that

$$(\nabla_X A)(Y,V) - (\nabla_Y A)(X,V) = (\nabla_X \nabla_Y dF)(V) - A(\nabla_X Y, V)$$
$$- (\nabla_Y \nabla_X dF)(V) + A(\nabla_Y X, V)$$
$$= (\nabla_X \nabla_Y dF)(V) - (\nabla_Y \nabla_X dF)(V)$$
$$- (\nabla_{\nabla_X Y - \nabla_Y X} dF)(V)$$
$$= (R^{F^* T\mathbb{R}^n \otimes T^* M}(X,Y)dF)(V)$$

because $\nabla_X Y - \nabla_Y X = [X,Y]$. To finalize the proof we calculate, using the fact that the Christoffel symbols in $(\mathbb{R}^n, \langle \cdot, \cdot \rangle)$ are all zero, in local coordinates

$$\nabla_{\frac{\partial}{\partial x^k}} \nabla_{\frac{\partial}{\partial x^l}} \left(dx^i \otimes \frac{\partial}{\partial y^\alpha} \right) \left(\frac{\partial}{\partial x^u} \right) =$$
$$= \nabla_{\frac{\partial}{\partial x^k}} \left(\nabla_{\frac{\partial}{\partial x^l}} dx^i \otimes \frac{\partial}{\partial y^\alpha} + dx^i \otimes \nabla_{\frac{\partial}{\partial x^l}} \frac{\partial}{\partial y^\alpha} \right) \left(\frac{\partial}{\partial x^u} \right)$$
$$= \left(\nabla_{\frac{\partial}{\partial x^k}} \nabla_{\frac{\partial}{\partial x^l}} dx^i \otimes \frac{\partial}{\partial y^\alpha} + \nabla_{\frac{\partial}{\partial x^l}} dx^i \otimes \nabla_{\frac{\partial}{\partial x^k}} \frac{\partial}{\partial y^\alpha} \right) \left(\frac{\partial}{\partial x^u} \right)$$
$$= \left(\nabla_{\frac{\partial}{\partial x^k}} \nabla_{\frac{\partial}{\partial x^l}} dx^i \otimes \frac{\partial}{\partial y^\alpha} \right) \left(\frac{\partial}{\partial x^u} \right)$$
$$= \left[\frac{\partial}{\partial x^k} \left(\frac{\partial}{\partial x^l} dx^i \left(\frac{\partial}{\partial x^u} \right) \right) - \frac{\partial}{\partial x^k} \left(dx^i \left(\nabla_{\frac{\partial}{\partial x^l}} \frac{\partial}{\partial x^u} \right) \right) \right.$$
$$\left. - \frac{\partial}{\partial x^l} \left(dx^i \left(\nabla_{\frac{\partial}{\partial x^k}} \frac{\partial}{\partial x^u} \right) \right) + dx^i \left(\nabla_{\frac{\partial}{\partial x^l}} \nabla_{\frac{\partial}{\partial x^k}} \frac{\partial}{\partial x^u} \right) \right] \frac{\partial}{\partial y^\alpha},$$

which implies that

$$\left(R^{F^*T\mathbb{R}^n\otimes T^*M}\left(\frac{\partial}{\partial x^k},\frac{\partial}{\partial x^l}\right)dF\right)\left(\frac{\partial}{\partial x^u}\right)$$
$$=\frac{\partial F^\alpha}{\partial x^i}\left[\nabla_{\frac{\partial}{\partial x^k}}\nabla_{\frac{\partial}{\partial x^l}}\left(dx^i\otimes\frac{\partial}{\partial y^\alpha}\right)-\nabla_{\frac{\partial}{\partial x^l}}\nabla_{\frac{\partial}{\partial x^k}}\left(dx^i\otimes\frac{\partial}{\partial y^\alpha}\right)\right]\left(\frac{\partial}{\partial x^u}\right)$$
$$=\frac{\partial F^\alpha}{\partial x^i}dx^i\left(\nabla_{\frac{\partial}{\partial x^l}}\nabla_{\frac{\partial}{\partial x^k}}\frac{\partial}{\partial x^u}-\nabla_{\frac{\partial}{\partial x^k}}\nabla_{\frac{\partial}{\partial x^l}}\frac{\partial}{\partial x^u}\right)\frac{\partial}{\partial y^\alpha}=-dF\left(R\left(\frac{\partial}{\partial x^k},\frac{\partial}{\partial x^l}\right)\frac{\partial}{\partial x^u}\right)$$

$\square$

We write the Codazzi equation in local coordinates as

$$\nabla_l A_{ij}-\nabla_i A_{lj}=R^k_{jli}F^\alpha_k$$

and with this we write the Codazzi equation considering A as a section in the normal bundle, $A\in\Gamma(TM^\perp\otimes TM^*\otimes TM^*)$,

$$\nabla^\perp_l A_{ij}-\nabla^\perp_i A_{lj}=[\nabla_l A_{ij}-\nabla_i A_{lj}]^\perp$$
$$\nabla^\perp_l A_{ij}-\nabla^\perp_i A_{lj}=[R^k_{jli}F^\alpha_k]^\perp=0 \quad (1.5)$$

The next important equation is the Gauß equation. For that we will need the following Lemma, where we understand $A\in\Gamma(F^*T\mathbb{R}^n\otimes TM^*\otimes TM^*)$.

Lemma 1.3.2. *Let (M,g) be a semi-Riemannian manifold, $F\colon (M,g)\to(\mathbb{R}^n,\langle\cdot,\cdot\rangle)$ be an isometric immersion, A be the second fundamental tensor of F and $X,Y,V,W\in\Gamma(TM)$. Then*

$$\langle(\nabla_X A)(Y,V),dF(W)\rangle=-\langle A(Y,V),A(X,W)\rangle. \quad (1.6)$$

Proof. Since $A(Y,V)_p\in T_pM^\perp$ for all $Y,V\in T_pM$, it follows that

$$0=X\langle A(Y,V),dF(W)\rangle=\langle\nabla_X(A(Y,V)),dF(W)\rangle+\langle A(Y,V),\nabla_X(dF(W))\rangle$$
$$=\langle(\nabla_X A)(Y,V)+A(\nabla_X Y,V)+A(Y,\nabla_X V),dF(W)\rangle$$
$$\quad+\langle A(Y,V),(\nabla_X dF)W+dF(\nabla_X W)\rangle$$
$$=\langle(\nabla_X A)(Y,V),dF(W)\rangle+\langle A(Y,V),(\nabla_X dF)W\rangle$$
$$=\langle(\nabla_X A)(Y,V),dF(W)\rangle+\langle A(Y,V),A(X,W)\rangle.$$

Proposition 1.3.3 (Gauß equation). *Let (M,g) be a semi-Riemannian manifold, $F:(M,g)\to(\mathbb{R}^n,\langle\cdot,\cdot\rangle)$ be an isometric immersion, $g:=F^*\langle\cdot,\cdot\rangle$ be the first fundamental form, A be the second fundamental tensor of F and $X,Y,V,W\in\Gamma(TM)$. Then*

$$\langle A(X,V),A(Y,W)\rangle-\langle A(Y,V),A(X,W)\rangle=-g(R(X,Y)V,W) \qquad (1.7)$$

Proof. It holds from the last Lemma that

$$\langle A(X,V),A(Y,W)\rangle-\langle A(Y,V),A(X,W)\rangle=$$
$$=\langle(\nabla_X A)(Y,V)-(\nabla_Y A)(X,V),dF(W)\rangle$$
$$=\langle\nabla_X(A(Y,V))-A(\nabla_X Y,V)-A(Y,\nabla_X V)-\nabla_Y(A(X,V)),dF(W)\rangle$$
$$+\langle A(\nabla_Y X,V)+A(X,\nabla_Y V),dF(W)\rangle$$
$$=\langle\nabla_X[(\nabla_Y dF)(V)]-[\nabla_{\nabla_X Y}dF](V)-[\nabla_Y dF](\nabla_X V),dF(W)\rangle$$
$$+\langle-\nabla_Y[(\nabla_X dF)(V)]+[\nabla_{\nabla_Y X}dF](V)+[\nabla_X dF](\nabla_Y V),dF(W)\rangle$$
$$=\langle\nabla_X[\nabla_Y dF(V)-dF(\nabla_Y V)]-\nabla_{\nabla_X Y}dF(V)+dF[\nabla_{\nabla_X Y}(V)],dF(W)\rangle$$
$$+\langle-\nabla_Y dF(\nabla_X V)+dF(\nabla_Y\nabla_X V)-\nabla_Y[\nabla_X dF(V)-dF(\nabla_X V)],dF(W)\rangle$$
$$+\langle\nabla_{\nabla_Y X}dF(V)-dF[\nabla_{\nabla_Y X}(V)]+\nabla_X dF(\nabla_Y V)-dF(\nabla_X\nabla_Y V),dF(W)\rangle$$
$$=-\langle dF[\nabla_X\nabla_Y V-\nabla_Y\nabla_X V-\nabla_{[X,Y]}V],dF(W)\rangle=-g(R(X,Y)V,W),$$

because $(\mathbb{R}^n,\langle\cdot,\cdot\rangle)$ has curvature 0:

$$0=\langle\nabla_X\nabla_Y dF(V)-\nabla_Y\nabla_X dF(V)-\nabla_{[X,Y]}dF(V).$$

□

We write this equation in local coordinates as

$$R_{klij}=\langle A_{ik},A_{jl}\rangle-\langle A_{jk},A_{il}\rangle. \qquad (1.8)$$

The next important equation is the Ricci-equation:

Proposition 1.3.4 (Ricci equation). *Let (M,g) be a semi-Riemannian manifold, $F:(M,g)\to(\mathbb{R}^n,\langle\cdot,\cdot\rangle)$ be an isometric immersion, A be the second fun-*

damental tensor of F, $R^\perp$ be the Riemannian curvature tensor of the normal bundle, $X, Y \in \Gamma(TM)$ and $\eta \in \Gamma(TM^\perp)$. Then

$$R^\perp(X,Y)\eta = \text{tr}(\langle \eta, A(Y, \cdot)\rangle A(X, \cdot)) - \text{tr}(\langle \eta, A(X, \cdot)\rangle A(Y, \cdot)). \tag{1.9}$$

Proof. Let us write $\{e_1 := \frac{\partial}{\partial x^1}, \ldots, e_m := \frac{\partial}{\partial x^m}\}$ for the local frame induced by the coordinates and $F_i := dF(e_i)$ for $i = 1, \ldots, m$. For $Z \in \Gamma(F^*T\mathbb{R}^n)$, it is not hard to see that the projection of Z in the normal bundle is given by the expression

$$Z^\perp = Z - \langle Z, F_i\rangle g^{il} F_l.$$

Using this and the fact that a section in $TM^\perp$ can be seen as a section in $F^*T\mathbb{R}^n$ we calculate

$$\begin{aligned}
\nabla_Y^\perp \eta &= \nabla_Y \eta - \langle \nabla_Y \eta, F_i\rangle g^{il} F_l = \nabla_Y \eta + \langle \eta, \nabla_Y F_i\rangle g^{il} F_l \\
&= \nabla_Y \eta + \langle \eta, \nabla_Y dF(e_i)\rangle g^{il} F_l \\
&= \nabla_Y \eta + \langle \eta, (\nabla_Y dF)(e_i) + dF(\nabla_Y e_i)\rangle g^{il} F_l \\
&= \nabla_Y \eta + \langle \eta, A(Y, e_i)\rangle g^{il} F_l.
\end{aligned}$$

From this follows

$$\begin{aligned}
\nabla_X^\perp \nabla_Y^\perp \eta &= \left[\nabla_X \left(\nabla_Y \eta + \left\langle \eta, A\left(Y, \frac{\partial}{\partial x^i}\right)\right\rangle g^{il} F_l\right)\right]^\perp \\
&= \left[\nabla_X \nabla_Y \eta + X\left(\langle \eta, A(Y, e_i)\rangle g^{il}\right) F_l + \langle \eta, A(Y, e_i)\rangle g^{il} \nabla_X F_l\right]^\perp \\
&= \left[\nabla_X \nabla_Y \eta + \langle \eta, A(Y, e_i)\rangle g^{il} \left((\nabla_X dF)(e_l) + dF(\nabla_X e_l)\right)\right]^\perp \\
&= [\nabla_X \nabla_Y \eta]^\perp + \langle \eta, A(Y, e_i)\rangle g^{il} A(X, e_l)
\end{aligned}$$

and finally

$$\begin{aligned}
R^\perp(X,Y)\eta &= \nabla_X^\perp \nabla_Y^\perp \eta - \nabla_Y^\perp \nabla_X^\perp \eta - \nabla_{[X,Y]}^\perp \eta \\
&= [\nabla_X \nabla_Y \eta - \nabla_Y \nabla_X \eta - \nabla_{[X,Y]} \eta]^\perp \\
&\quad + \langle \eta, A(Y, e_i)\rangle g^{il} A(X, e_l) - \langle \eta, A(X, e_i)\rangle g^{il} A(Y, e_l) \\
&= \langle \eta, A(Y, e_i)\rangle g^{il} A(X, e_l) - \langle \eta, A(X, e_i)\rangle g^{il} A(Y, e_l),
\end{aligned}$$

because the curvature of the bundle $F^*T\mathbb{R}^n$ is zero. $\square$

The Riemannian curvature tensor of the normal bundle $R^\perp_{ij}$ can be seen as the section $\langle R^\perp(\frac{\partial}{\partial x^i}, \frac{\partial}{\partial x^j})\cdot,\cdot\rangle \in \Gamma(T^*M^\perp \otimes T^*M^\perp)$. To do this we identify the bundle $(TM^\perp)^*$ with the bundle $TM^\perp$ through the inner product, so that we can consider the Riemannian curvature in the normal bundle as a 2-form (from the definition it is clear the anti-symmetry in the two vectorfields over TM) over M with values in $TM^\perp \otimes TM^\perp$ and denote it with $R^\perp_{ij}$. In this notation the Ricci equation is written

$$R^\perp_{ij} = A_{jk} \otimes A^k_i - A_{ik} \otimes A^k_j =: A_{jk} \wedge A^k_i. \tag{1.10}$$

We end this section with the very usefull commutation formula:

Lemma 1.3.5. *Let M be a m-dimensional differentiable manifold, E be a vector bundle of dimension δ over M and ∇ be a connection on this bundle. If $T^{\alpha_1...\alpha_\phi}_{k_1...k_r} \in \Gamma(T^*M \otimes ... \otimes T^*M \otimes E \otimes ... \otimes E)$ then*

$$\begin{aligned}\nabla_i \nabla_j T^{\alpha_1...\alpha_\phi}_{k_1...k_r} - \nabla_j \nabla_i T^{\alpha_1...\alpha_\phi}_{k_1...k_r} =& -\sum_{h=1}^r \sum_{p=1}^m R^p_{k_h ij} T^{\alpha_1...\alpha_\phi}_{k_1...k_{h-1}pk_{h+1}...k_r} \\ &+ \sum_{h=1}^\phi \sum_{\beta=1}^\delta R^{\alpha_h}_{\beta ij} T^{\alpha_1...\alpha_{h-1}\beta\alpha_{h+1}...\alpha_\phi}_{k_1...k_r},\end{aligned} \tag{1.11}$$

Proof. We prove by induction. To make the equations smaller let us write $e_i := \frac{\partial}{\partial x^i}$ and $\{E_1,...,E_\delta\}$ for a local trivialization of E and E^*_α to the dual of E_α. Let $X \in \Gamma(E)$ with $X = X^\alpha E_\alpha$.

$$\nabla_i \nabla_j X = (\nabla_{e_i}(\nabla X))(e_j) = \nabla_{e_i}\nabla_{e_j}X - \nabla_{\nabla_{e_i}e_j}X$$

which implies that

$$\begin{aligned}\nabla_i \nabla_j X - \nabla_j \nabla_i X =& \nabla_{e_i}\nabla_{e_j}X - \nabla_{\nabla_{e_i}e_j}X - \nabla_{e_j}\nabla_{e_i}X + \nabla_{\nabla_{e_j}e_i}X \\ =& \nabla_{e_i}\nabla_{e_j}X - \nabla_{e_j}\nabla_{e_i}X - \nabla_{\nabla_{e_i}e_j - \nabla_{e_j}e_i}X = R(e_i,e_j)X\end{aligned}$$

because the Levi-Civita connection is torsionfree. The case $Y \in \Gamma(T^*M)$ is analogous.

Let us now assume that the Theorem is true for some ϕ and r fixed. Then for $\phi+1$ we get

$$\nabla_i \nabla_j (T^{\alpha_1...\alpha_{\phi+1}}_{k_1...k_r} E_{\alpha_1} \otimes ... \otimes E_{\alpha_{\phi+1}} \otimes dx^{k_1} \otimes ... \otimes dx^{k_r})$$
$$- \nabla_j \nabla_i (T^{\alpha_1...\alpha_{\phi+1}}_{k_1...k_r} E_{\alpha_1} \otimes ... \otimes E_{\alpha_{\phi+1}} \otimes dx^{k_1} \otimes ... \otimes dx^{k_r})$$

$$= \nabla_i \nabla_j (T^{\alpha_1...\alpha_{\phi+1}}_{k_1...k_r} E_{\alpha_1} \otimes ... \otimes E_{\alpha_\phi} \otimes dx^{k_1} \otimes ... \otimes dx^{k_r}) \otimes E_{\phi+1}$$
$$- \nabla_j \nabla_i (T^{\alpha_1...\alpha_{\phi+1}}_{k_1...k_r} E_{\alpha_1} \otimes ... \otimes E_{\alpha_\phi} \otimes dx^{k_1} \otimes ... \otimes dx^{k_r}) \otimes E_{\phi+1}$$
$$+ T^{\alpha_1...\alpha_\phi}_{k_1...k_r} E_{\alpha_1} \otimes ... \otimes E_{\alpha_\phi} \otimes dx^{k_1} \otimes ... \otimes dx^{k_r} \otimes (\nabla_i \nabla_j E_{\phi+1} - \nabla_j \nabla_i E_{\phi+1})$$
$$= -\sum_{h=1}^{r}\sum_{p=1}^{m} R^p_{k_h ij} T^{\alpha_1...\alpha_{\phi+1}}_{k_1...k_{h-1}pk_{h+1}...k_r} E_{\alpha_1} \otimes ... \otimes E_{\alpha_{\phi+1}} \otimes dx^{k_1} \otimes ... \otimes dx^{k_r}$$
$$+ \sum_{h=1}^{\phi}\sum_{\beta=1}^{\delta} R^{\alpha_h}_{\beta ij} T^{\alpha_1...\alpha_{h-1}\beta\alpha_{h+1}...\alpha_{\phi+1}}_{k_1...k_r} E_{\alpha_1} \otimes ... \otimes E_{\alpha_{\phi+1}} \otimes dx^{k_1} \otimes ... \otimes dx^{k_r}$$
$$+ \sum_{\beta=1}^{\delta} T^{\alpha_1...\alpha_\phi\beta}_{k_1...k_r} \delta R^{\phi+1}_{\beta ij} E_{\alpha_1} \otimes ... \otimes E_{\alpha_{\phi+1}} \otimes dx^{k_1} \otimes ... \otimes dx^{k_r}$$
$$= -\sum_{h=1}^{r}\sum_{p=1}^{m} R^p_{k_h ij} T^{\alpha_1...\alpha_{\phi+1}}_{k_1...k_{h-1}pk_{h+1}...k_r} E_{\alpha_1} \otimes ... \otimes E_{\alpha_{\phi+1}} \otimes dx^{k_1} \otimes ... \otimes dx^{k_r}$$
$$+ \sum_{h=1}^{\phi+1}\sum_{\beta=1}^{\delta} R^{\alpha_h}_{\beta ij} T^{\alpha_1...\alpha_{h-1}\beta\alpha_{h+1}...\alpha_{\phi+1}}_{k_1...k_r} E_{\alpha_1} \otimes ... \otimes E_{\alpha_{\phi+1}} \otimes dx^{k_1} \otimes ... \otimes dx^{k_r},$$

where we wrote the sum on the indices h, p and β because it might not be clear up to which natural number they are summing, since it is not always up to the maximal possible. We took $E_{\alpha_{\phi+1}}$ apart just to mean that it was not to be considered in the covariant derivatives, but, if one changes the order in such a tensor product, it is also necessary to change the order of the components in the resulting tensor to compare it with the original one. This completes the proof for $\phi+1$. The case $r+1$ is analogous. $\square$

Chapter 2

Hyperquadric Homotheties of the MCF

We are interested in the homotheties generated by the mean curvature flow in $(\mathbb{R}^n, \langle \cdot, \cdot \rangle)$. There are two types of homotheties, the ones that shrink (called self-shrinkers) and the ones that grow (called self-expanders)[1]. In the Euclidean case the mean curvature flow (MCF) has no compact self-expanders because the comparison with Spheres (Proposition 3.3 of [Eck04]) shows that a compact initial immersion will stay inside any sphere that contains it, and thence cannot grow. There are many results about self-shrinkers in the Euclidean case. For example in [AL86], U. Abresch & J. Langer found the self-shrinker curves on the plane; in [Hui93], G. Huisken classified the self-shrinking hypersurfaces with non-negative (scalar) mean curvature and in [Smo05], K. Smoczyk classified the self-shrinkers in higher codimension with principal normal parallel in the normal bundle. But even in the Riemannian case there is no complete classification of all possible homotheties generated by the MCF, so that finding such a classification in a pseudo-euclidean space would be an unattainable objective. In this chapter we consider the hyperquadric homotheties of the MCF.

2.1 Hyperquadric Homotheties of the MCF

We now consider surfaces that change homothetically in time with the MCF. Without loss of generality let the homothety center be the origin of $(\mathbb{R}^n, \langle \cdot, \cdot \rangle)$.

---

[1] The stationary solutions are not considered in this work because we need $\|\vec{H}\|^2 \neq 0$.

So we consider that there is a rescaling function $c(t)$, such that the rescaled immersion does not change the surface (as a set).

Definition 2.1.1. Let M be a smooth manifold, (N,h) be a semi-Riemannian manifold and $F_0 : M \to N$ be an immersion. A smooth a family of isometric immersions $F : M \times [0,T) \to N$, for some $T > 0$, such that the metric $g_t := F(\cdot,t)^*h$ is nondegenerate for all $t \in [0,T)$ is called a *solution of the mean curvature flow with initial immersion* F_0 if it satisfies

$$\frac{dF}{dt}(p,t) = \vec{H}(p), \quad \text{and} \quad F(p,0) = F_0(p) \; \forall p \in M, t \in [0,T), \quad (2.1)$$

where $\vec{H}$ is the mean curvature vector of the immersion $F(\cdot,t) : M \to (N,h)$.

Now we consider properties of homotheties generated by the mean curvature flow. Let $F : M \times [0,T) \to (\mathbb{R}^n, \langle \cdot, \cdot \rangle)$ be a solution of the MCF for some initial immersion. Suppose that this solution generates a homothety, this means that there is a differentiable rescaling function $c : [0,T) \to (0,\infty)$ with $c(0) = 1$ such that the rescaled immersion only changes $F(M)$ in tangential directions and thence preserves the form of the initial surface, so that $\tilde{F} := cF$ satisfies

$$\left\langle \frac{d\tilde{F}}{dt}(p,t), \tilde{\nu} \right\rangle = 0 \qquad \forall \tilde{\nu} \in T_p M^\perp, \quad (2.2)$$

where $T_p M^\perp$ is the normal space of M at point p with respect to the immersion $\tilde{F}(\cdot,t)$ (which is the same as the normal space of $F(\cdot,t)$ because $\tilde{F}$ is just a rescaling of F.) and the following holds:

$$0 = \left\langle \frac{d\tilde{F}}{dt}, \tilde{\nu} \right\rangle = \left\langle c\frac{dF}{dt} + \dot{c}F, \tilde{\nu} \right\rangle \qquad \forall \tilde{\nu} \in TM^\perp,$$

which implies

$$\vec{H} = -\frac{\dot{c}}{c} F^\perp.$$

Definition 2.1.2. Let $F : M \times [0,T) \to (\mathbb{R}^n, \langle \cdot, \cdot \rangle)$ be a solution of the MCF for some initial immersion. If there is a differentiable function $c : [0,T) \to (0,\infty)$ with $c(0) = 1$ such that

$$\vec{H} = -\frac{\dot{c}}{c} F^\perp, \quad (2.3)$$

we say that F is an homothety of the MCF.

Where we used the fact that the mean curvature vector $\vec{H}(p,t)$ is normal (with respect to the inner product $\langle\cdot,\cdot\rangle$ of $\mathbb{R}^n$) to T_pM $\forall (p,t) \in M \times [0,T)$. Beyond this, as the Christoffel symbols of $(\mathbb{R}^n, \langle\cdot,\cdot\rangle)$ are all zero, we have that $\vec{H}(t) = \nabla^j \nabla_j F(t) = \triangle F(t)$, where $\triangle F$ is the Laplace-Beltrami[2] operator of (M, g_t) on real functions of M applied to each component F^α of F.

From now on, we use normal, norm, inner product and others always with respect to the pseudo-euclidean inner product, unless we say the contrary. We will not use the index $_t$ on the metric anymore, just indices for its representation in local coordinates, although the metric always depends on t. We shall also write $F(t)$ for $F(\cdot, t)$ when whishing to make explicit mention to the time t.

In particular we look now at the hyperquadric homotheties of the MCF in $(\mathbb{R}^n, \langle\cdot,\cdot\rangle)$

Definition 2.1.3. Let $G: M \to (\mathbb{R}^n, \langle\cdot,\cdot\rangle)$ be an immersion. We say that G *lies in some hyperquadric* or *is hyperquadric* if

$$\|G\|^2 = k,$$

$k \in \mathbb{R}$ constant, for all $x \in M$.

From now on let $F: M \times [0, T) \to (\mathbb{R}^n, \langle\cdot,\cdot\rangle)$ be a homothety of the MCF. It could happen that some solutions of the flow in which the initial immersion lies in a hyperquadric, i. e. $F(x, 0) \subset \mathcal{H}^{n-1}(k)$ for all $x \in M$, cease lying in some hyperquadric during the flow, i. e. there is no function $k(t)$ such that $F(x, t) \subset \mathcal{H}^{n-1}(k(t))$ for all $x \in M$. This cannot happen for homotheties, as the following result states:

Lemma 2.1.4. *If $F(0, x) \subset \mathcal{H}^{n-1}(k(0))$ for all $x \in M$ then $F(t, x) \subset \mathcal{H}^{n-1}(k(t))$ for all $x \in M$, for some function $k: [0, T) \to \mathbb{R}$.*

Proof. From equation (2.2) we know that the rescaled immersion $\tilde{F}$ is such that $\frac{d\tilde{F}}{dt} \in \Gamma(TM)$ for all $t \in [0, T)$, this means that $\tilde{F}$ just moves points over the initial surface, this implies $\|\tilde{F}\|^2 = k(0)$ for all $t \in [0, T)$. So that $\tilde{F} = c(t)F$ implies that $\|F(t, x)\|^2 = c^{-2}(t)k(0)$, and as this does not depend on the point $x \in M$, so it is in some hyperquadric. □

---
[2] If the metric is positive-definite, which is mainly the case in this work, but the results for hyperquadric solutions in this chapter do not depend on the signature of the metric.

Definition 2.1.5. A *hyperquadric homothety of the mean curvature flow* $F : M \times [0,T) \to (\mathbb{R}^n, \langle \cdot, \cdot \rangle)$ is a homothety of the mean curvature flow, such that $F(\cdot, t)$ is hyperquadric for all $t \in [0,T)$.

As the position vector in a hyperquadric is normal (with respect to the inner product that generates the hyperquadric) to the hyperquadric, it follows that F is always orthogonal to $\mathcal{H}^{n-1}(k(t))$:

Corollary 2.1.6. $F(x,t) \in T_x M^\perp$ for all $(x,t) \in M \times [0,T)$.

Proof. In local coordinates, for any $t \in [0,T)$

$$0 = \frac{\partial}{\partial x^l}[c^{-2}(t)k(0)] = \nabla_l \|F(t)\|^2 = 2\langle F_l(t), F(t)\rangle \ \forall l \in \{1,\ldots,m\}$$

which implies that $F(t)$ stays in $F(M,t)^\perp$ as long as a solution exists. $\square$

With this, $\|F\|^2$ can be calculated:

Lemma 2.1.7. $\|F(t)\|^2 = k(0) - 2mt$ for $t \in [0,T)$.

Proof. For $t = 0$ it is clear that $\|F(0)\|^2 = k(0)$. And for all $t \in [0,T)$:

$$\frac{d}{dt}\|F(t)\|^2 = 2\left\langle \frac{d}{dt}F(t), F(t)\right\rangle = 2\left\langle \vec{H}(t), F(t)\right\rangle$$
$$= 2\left\langle \nabla^j \nabla_j F(t), F(t)\right\rangle = -2\langle F(t)_j, F(t)_l\rangle g^{jl} = -2m,$$

where Corollary 2.1.6 is used for $\nabla^j \langle F_j, F\rangle = 0$, this completes the proof. $\square$

Definition 2.1.8. Let (M,g) and (N,h) be semi-Riemannian manifolds and $G : M \to N$ an isometric immersion. G is called a *minimal immersion* if the mean curvature vector $\vec{H}_G$ of this immersion is identically zero

$$\vec{H}_G = 0.$$

If G is a minimal immersion we say that M (or $G(M)$) is a *minimal surface* of N.

We use the name *minimal* because the condition $\vec{H} = 0$ is then mnemonic, but the "minimal immersions" in our sense are just critical points of the volume functional, not always minima of this functional, as it occurs in the euclidean case.

We prove now that a hyperquadric homothety of the mean curvature flow is a minimal immersion in the hyperquadric $\mathcal{H}^{n-1}(k(0)-2mt)$ for all $t \in [0,T)$. We will need the following Lemma:

Lemma 2.1.9. *Let $F: M \to N$ and $G: N \to P$ be isometric immersions between semi-Riemannian manifolds $(M,g), (N,h)$ and (P,l). Denote $\vec{H}_F$, $\vec{H}_G$ and $\vec{H}_{G \circ F}$ the mean curvatures of F, G and $G \circ F$ respectively. Then:*

$$\vec{H}_{G \circ F} = dG(\vec{H}_F) + \mathrm{tr}_M(\nabla dG)(dF\cdot, dF\cdot),$$

where tr_M is the trace with respect to (M,g), given in local coordinates by

$$\mathrm{tr}_M(\nabla dG)(dF\cdot, dF\cdot) = g^{ij}(\nabla dG)\left(dF\left(\frac{\partial}{\partial x^i}\right), dF\left(\frac{\partial}{\partial x^j}\right)\right).$$

Proof. For any $x \in M$ it holds

$$(\nabla d(G \circ F))_x = \nabla(dG_{F(x)} \circ dF)$$
$$(\nabla d(G \circ F))_x = (\nabla dG)_{F(x)}(dF\cdot, dF\cdot) + dG_{F(x)} \circ \nabla dF, \qquad (2.4)$$

because the Levi-Civita connection of M is just the tangent projection of the Levi-Civita connection of N and this product rule can be calculated as in Lemma 1.2.7 with different bundles. The result follows if we take the trace. □

Then it can be proven:

Theorem 2.1.10. *Let $F: M \times [0,T) \to (\mathbb{R}^n, \langle \cdot, \cdot \rangle)$ be a hyperquadric homothety of the mean curvature flow, then $F(M,t)$ is a minimal surface of the hyperquadric $\mathcal{H}^{n-1}(\|F(0)\|^2 - 2mt)$ for all $t \in [0,T)$.*

Proof. We consider the natural inclusion $I(t)$ of the hyperquadric $\mathcal{H}^{n-1}(k(0)-2mt)$ into $(\mathbb{R}^n, \langle \cdot, \cdot \rangle)$ [3], which is a diffeomorphism on its image, and the immersion $G: M \to \mathcal{H}^{n-1}(k)$ defined as $G := I^{-1} \circ F$, as in the figure:

---

[3] If $k(0) - 2mt = 0$ we consider $I(t) : \mathcal{H}^{n-1}(0) \setminus \{0\} \to \mathbb{R}^n$ because $\mathcal{H}^{n-1}(0)$ is not an $n-1$-dimensional manifold. But even in some generalized sense the point 0 would not be a great trouble for a homothety because, if $F(x,t) = 0$, then $\vec{H}(x,t) = \dot{c}/cF^{\perp}(x,t) = 0$, and the minimal condition would be held at this point.

$$\mathcal{H}^{n-1}(k(0)-2mt) \xrightarrow{I} (\mathbb{R}^n, \langle \cdot, \cdot \rangle)$$
$$G \uparrow \quad \nearrow F$$
$$M$$

Let us write $\vec{H}_F$, $\vec{H}_G$ $\vec{H}_I$ for the mean curvature vectors of F, G and I respectively. About the terms in the relation between $\vec{H}_F$ and $\vec{H}_G$ given by Lemma 2.1.9 the following holds

- $\vec{H}_F \in TM^\perp$,
- $dI(\vec{H}_G) \in (dI(T\mathcal{H}^{n-1}))$,
- $g^{ij}(\nabla dI)\left(dG\left(\frac{\partial}{\partial x^i}\right), dG\left(\frac{\partial}{\partial x^j}\right)\right) \in T\mathcal{H}^{n-1\perp}$ for this is just a sum on the second fundamental tensor of the immersion I.

If F is a hyperquadric homothety of the MCF, equation (2.3) implies that there is a function $\beta : [0,T) \to \mathbb{R}$ such that

$$\vec{H}_F = \beta(t) F^\perp = \beta(t) F \implies \vec{H}_F \in (T\mathcal{H}^{n-1})^\perp.$$

Thus, $dI(\vec{H}_G)$ is the only term tangential to the hyperquadric in Lemma 2.1.9, thence $dI(\vec{H}_G) = 0$ and $\vec{H}_G = 0$ (for I is an immersion). $\square$

Further, we can calculate β. Let $t \in [0,T)$ be fixed and $x \in M$ be any point

$$\beta \|F(x,t)\|^2 = \langle \vec{H}_F, F \rangle = -\langle F_i, F_j \rangle g^{ij} = -m$$
$$\implies \beta(t) = -\frac{m}{\|F(t)\|^2}, \tag{2.5}$$

and Lemma 2.1.7 implies that

$$\vec{H}_F(t) = -\frac{m}{\|F(0)\|^2 - 2mt} F(t). \tag{2.6}$$

2.2 Existence and Uniqueness

2.2.1 Immersion in the Hyperquadric $\mathcal{H}^{n-1}(k)$ with $k > 0$.

Let $\|F(x,0)\|^2 = k > 0$ for all $x \in M$. To found the $F(x,t)$ explicitly we need, from eq. (2.3) and (2.6), to solve the ordinary differential equation

$$\frac{\dot{c}}{c} = \frac{m}{k - 2mt},$$

with initial condition $c(0) = 1$. This equation has a unique solution, which is the function $c : [0, T) \to \mathbb{R}$, with $T = \frac{k}{2m}$, defined through

$$c(t) := \sqrt{k}(k - 2mt)^{-1/2},$$

so that

$$\dot{c} := \frac{dc}{dt} = m\sqrt{k}(k - 2mt)^{-3/2}.$$

It follows from equation (2.6), for any $(x,t) \in M \times [0,T)$ that

$$\frac{d}{dt}F(x,t) = \vec{H}_{F(\cdot,t)}(x) = -\frac{m}{k-2mt}F(x,t) = -\frac{\dot{c}}{c}F(x,t) \Longrightarrow \frac{d}{dt}(cF(t)) = 0. \quad (2.7)$$

Hence $cF(x,t) = F(x,0)$ and

$$F(x,t) = \frac{1}{c}F(x,0). \quad (2.8)$$

By construction we proved that if there is a hyperquadric homothety of the MCF, then every other solution of the MCF has to be equal to it, so it is unique. The pull back $F^*\langle \cdot, \cdot \rangle$ stays invertible all along the flow (for $g_{ij}(t) = \left(\frac{1}{c(t)}\right)^2 g_{ij}(0)$) so that these constructions make sense for $t \in [0,T)$. It is clear that T is the maximal time existence, because $\lim_{t \to T} F(x,t) = 0$ for all $x \in M$. So, if there is a solution of the mean curvature flow, this will be only one.

We still have to deal with the question of existence. As we saw in Theorem 2.1.10, a hyperquadric homothety of the mean curvature flow has to be a minimal surface of the hyperquadric. This motivates the following Theorem:

Theorem 2.2.1. *Let $F : M^m \to (\mathbb{R}^n, \langle \cdot, \cdot \rangle)$ be an immersion such that $g := F^*\langle \cdot, \cdot \rangle$ is nondegenerate and $\|F\|^2 = k \in \mathbb{R}$, $k > 0$, then the solution of the MCF* of this initial immersion is a homothety if, and only if, $F: M \to \mathcal{H}^{n-1}(k)$ is a minimal immersion in the hyperquadric $\mathcal{H}^{n-1}(k)$. The mean curvature flow of F has a solution $F: M \times [0,T) \to (\mathbb{R}^n, \langle \cdot, \cdot \rangle)$ with $T = \frac{k}{2m}$; moreover, the solution is $F(x,t) := c^{-1}(t)F(x)$, with $c(t) := \sqrt{k}(k - 2mt)^{-1/2}$, $\forall (x,t) \in M \times [0,T)$.

Proof. We have already proven that a homothety with initial immersion in a hyperquadric is a minimal immersion in some hyperquadric for every $t \in [0,T)$ (Theorem 2.1.10). We now need to prove that the MCF of a minimal immersion generates a homothety.

Let us write $F(t) := F(\cdot, t)$. We consider the natural inclusion I of the hyperquadric $\mathcal{H}^{n-1}(k)$ into $(\mathbb{R}^n, \langle \cdot, \cdot \rangle)$ and the immersion $G: M \to \mathcal{H}^{n-1}(k)$ defined as $G := I^{-1} \circ F$. By Lemma 2.1.9 it is true that

$$\vec{H}_{F(0)} = \vec{H}_G + g^{ij}(\nabla dI)\left(dG\left(\frac{\partial}{\partial x^i}\right), dG\left(\frac{\partial}{\partial x^j}\right)\right)$$
$$= g^{ij}(\nabla dI)\left(dG\left(\frac{\partial}{\partial x^i}\right), dG\left(\frac{\partial}{\partial x^j}\right)\right),$$

because $F(0)$ is a minimal immersion on the hyperquadric. Moreover, $\vec{H}_{F(0)}$ is orthogonal to $\mathcal{H}^{n-1}(\|F(0)\|^2)$ (for the second fundamental form of I is orthogonal to $\mathcal{H}^{n-1}(\|F(0)\|^2)$), but so is $F(0)$, such that there is a function $\varphi: M \to \mathbb{R}$ with $\vec{H}_{F(0)} = \varphi F(0)$ (for the codimension of the hyperquadric is one). One can calculate φ:

$$\varphi \|F(0)\|^2 = \langle \vec{H}_{F(0)}, F(0) \rangle = -g^{ij}\langle \nabla_j F(0), \nabla_i F(0) \rangle = -m$$
$$\implies \vec{H}_{F(0)} = -\frac{m}{\|F(0)\|^2}F(0).$$

Now consider the following deformation: $F(t) = c^{-1}(t)F(0)$, which is an homothety by definition; then it holds $g^{ij}(t) = c^2(t)g^{ij}(0)$. But $\vec{H}_{F(t)} = g^{ij}\nabla_i\nabla_j F(t)$ implies

$$\vec{H}_{F(t)} = c(t)\vec{H}_{F(0)} = -\frac{m}{\|F(0)\|^2}c(t)F(0).$$

On the other hand, for the function c given in the statement of the Theorem,

$$\begin{aligned}\frac{dF(t)}{dt} &= \frac{d}{dt}\left(\frac{1}{c(t)}\right)F(0) = -\frac{\dot{c}}{c^2}(t)F(0)\\ &= -\frac{m\sqrt{\|F(0)\|^2}(\|F(0)\|^2 - 2mt)^{-3/2}}{\|F(0)\|^2(\|F(0)\|^2 - 2mt)^{-1}}F(0)\\ &= -\frac{m}{\|F(0)\|^2}c(t)F(0) = \vec{H}_{F(t)}.\end{aligned}$$

Therefore this is a solution of the mean curvature flow. That this is the only solution in the class of homotheties follows from the explicit solution (eq. (2.8)) previously given. □

2.2.2 Immersion in the Hyperquadric $\mathcal{H}^{n-1}(k)$ with $k < 0$.

Let $\|F(x,0)\|^2 = k < 0$ for all $x \in M$. To found the $F(x,t)$ explicitly we need, from eq. (2.3) and (2.6), to solve the ordinary differential equation

$$\frac{\dot{c}}{c} = \frac{m}{k - 2mt},$$

with initial condition $c(0) = 1$. This equation has a unique solution, which is the function $c : [0, \infty) \to \mathbb{R}$, defined through

$$c(t) := \sqrt{-k}(-k + 2mt)^{-1/2},$$

so that

$$\dot{c} := \frac{dc}{dt} = -m\sqrt{-k}(-k + 2mt)^{-3/2}.$$

It follows from equation (2.6), for any $(x,t) \in M \times [0,T)$, that

$$\frac{d}{dt}F(x,t) = \vec{H}_F(\cdot,t)(x) = -\frac{m}{k - 2mt}F(x,t) = -\frac{\dot{c}}{c}F(x,t) \implies \frac{d}{dt}(cF(x,t)) = 0. \tag{2.9}$$

Hence is $cF(x,t) = F(x,0)$ and

$$F(x,t) = \frac{1}{c}F(x,0). \tag{2.10}$$

By construction we proved that if there is a hyperquadric homothety of the

MCF, then every other solution has to be equal to it, so it is unique. The pull back $F^*\langle\cdot,\cdot\rangle$ stays invertible all along the flow (for $g_{ij}(t) = \left(\frac{1}{c(t)}\right)^2 g_{ij}(0)$) so that these constructions make sense for $t \in [0,\infty)$ and we have long time existence, moreover, if there is a solution of the mean curvature flow, this is the only one.

We still have to deal with the question of existence. As we saw in Theorem 2.1.10, a hyperquadric homothety of the mean curvature flow has to be a minimal surface of the hyperquadric. This motivates the following Theorem:

Theorem 2.2.2. *Let $F : M^m \to (\mathbb{R}^n, \langle\cdot,\cdot\rangle)$ be an immersion such that $g := F^*\langle\cdot,\cdot\rangle$ is nondegenerate and $\|F\|^2 = k \in \mathbb{R}$, $k < 0$, then the solution of the MCF of this initial immersion is a homothety if, and only if, $F : M \to \mathcal{H}^{n-1}(k)$ is a minimal immersion in the hyperquadric $\mathcal{H}^{n-1}(k)$. The mean curvature flow of F has a solution $F(t) : M \times [0,\infty) \to (\mathbb{R}^n, \langle\cdot,\cdot\rangle)$; moreover, the solution is $F(x,t) := c^{-1}(t)F(x)$, with $c(t) := \sqrt{-k}(-k + 2mt)^{-1/2}$, for all $(x,t) \in M \times [0,\infty)$.*

Proof. We have already proven that a homothety with initial immersion in a hyperquadric is a minimal immersion in some hyperquadric for every $t \in [0,T)$ (Theorem 2.1.10). We now need to prove that the MCF of a minimal immersion generates a homothety.

Let us write $F(t) := F(\cdot, t)$. We consider the natural inclusion I of the hyperquadric $\mathcal{H}^{n-1}(k)$ into $(\mathbb{R}^n, \langle\cdot,\cdot\rangle)$ and the immersion $G : M \to \mathcal{H}^{n-1}(k)$ defined as $G := I^{-1} \circ F$. By Lemma 2.1.9 it is true that

$$\vec{H}_{F(0)} = \vec{H}_G + g^{ij}(\nabla dI)\left(dG\left(\frac{\partial}{\partial x^i}\right), dG\left(\frac{\partial}{\partial x^j}\right)\right)$$
$$= g^{ij}(\nabla dI)\left(dG\left(\frac{\partial}{\partial x^i}\right), dG\left(\frac{\partial}{\partial x^j}\right)\right),$$

because $F(0)$ is a minimal immersion on the hyperquadric. Moreover, $\vec{H}_{F(0)}$ is orthogonal to $\mathcal{H}^{n-1}(\|F(0)\|^2)$ (for the second fundamental form of I is orthogonal to $\mathcal{H}^{n-1}(\|F(0)\|^2)$), but so is $F(0)$, such that there is a function $\varphi : M \to \mathbb{R}$ with $\vec{H}_{F(0)} = \varphi F$ (for the codimension of the hyperquadric is one). One can calculate φ:

$$\varphi\|F(0)\|^2 = \langle \vec{H}_{F(0)}, F(0)\rangle = -g^{ij}\langle\nabla_j F(0), \nabla_i F(0)\rangle = -m$$
$$\implies \vec{H}_{F(0)} = -\frac{m}{\|F(0)\|^2}F(0).$$

Now consider the following deformation: $F(t) = c^{-1}(t)F(0)$, which is an homothety by definition; then it holds $g^{ij}(t) = c^2(t)g^{ij}(0)$. But $\vec{H}_{F(t)} = g^{ij}\nabla_i\nabla_j F(t)$ implies

$$\vec{H}_{F(t)} = c(t)\vec{H}_{F(0)} = -\frac{m}{\|F(0)\|^2}c(t)F(0).$$

On the other hand, for the function c given in the statement of the Theorem,

$$\begin{aligned}\frac{dF(t)}{dt} &= \frac{d}{dt}\left(\frac{1}{c(t)}\right)F(0) = -\frac{\dot{c}}{c^2}(t)F(0)\\ &= \frac{m\sqrt{-\|F(0)\|^2}(-\|F(0)\|^2 + 2mt)^{-3/2}}{-\|F(0)\|^2(-\|F(0)\|^2 + 2mt)^{-1}}F(0)\\ &= -\frac{m}{\|F(0)\|^2}c(t)F(0) = \vec{H}_{F(t)}.\end{aligned}$$

Therefore this is a solution of the mean curvature flow. That this is the only solution in the class of homotheties follows from the explicit solution (eq. (2.10)) previously given. □

2.2.3 Immersion in the Hyperquadric $\mathcal{H}^{n-1}(0)$

Let $F : M \times [0,T)$ be a homothety generated by the MCF with $\|F(x,0)\|^2 = 0$ for all $x \in M$. From Lemma 2.1.7 it holds $\|F(x,t)\|^2 = -2mt$ if $F^*\langle\cdot,\cdot\rangle$ is nondegenerate, so that

$$\|F(x,t)\|^2 < 0 \tag{2.11}$$

for all $(x,t) \in M \times (0,T)$.

On the other hand, $c(t)F(x,t) \in F(M,0)$ because F is a homothety, so that

$$0 = \|c(t)F(x,t)\|^2 = c(t)^2\|F(x,t)\|^2$$

But $c(t) \neq 0$ because $F(M,t) = \{0\}$ for $c(t) = 0$, which cannot be an immersion, then $\|F(x,t)\|^2 = 0$ for all $t \in [0,T)$. Which is a contradiction to eq. (2.11). So we proved

Theorem 2.2.3. *There are no hyperquadric homotheties of the MCF $F : M \times [0,T) \to (\mathbb{R}^n, \langle\cdot,\cdot\rangle)$ with nondegenerate metric such that $F(M,0) \subset \mathcal{H}^{n-1}(0)$.*

□

Remark 2.2.4. One could expect to find at least some stationary solutions in the light cone, like straight lines, but for such a line the metric is degenerate and thence this case is not included in Theorem 2.2.3.

Remark 2.2.5. But, as Ecker noted in [Eck97], the upper light-cone would immediately change to a hyperquadric and the explicitly solution to the MCF with the upper light-cone as initial condition in $\mathbb{R}^{1,n}$ is given by the family of graphs $\delta_t : \mathbb{R}^{n-1} \to \mathbb{R}$
$$\delta_t(x) = \sqrt{\|x\|_{\mathbb{E}}^2 + 2(n-1)t},$$
for any $t \in [0, \infty)$, which is a homothety after $t = 0$.

Remark 2.2.6. There is an important difference between minimal immersions in $\mathcal{H}^{n-1}(k)$ for $k > 0$ and $k < 0$.

If $\|F(x,0)\|^2 = k > 0$, for $x \in M$, then
$$\vec{H}_{F(\cdot,t)}(x) = -\frac{m}{k}c^2(t)F(x,t)$$
has the opposite sign as $F(x,t)$, and F shrinks under the MCF.

If $\|F(x,0)\|^2 = k < 0$, for $x \in M$, then
$$\vec{H}_{F(\cdot,t)}(x) = -\frac{m}{k}c^2(t)F(x,t)$$
has the same sign as $F(x,t)$, and F expands under the MCF.

Definition 2.2.7. Let (N, h) be a semi-Riemannian manifold, M a smooth manifold and $F : M \to N$ an immersion such that the mean curvature vector of F satisfies $\|\vec{H}(x)\|^2 \neq 0$ for all $x \in M$. The *principal normal* is the vectorfield
$$\nu := \frac{\vec{H}}{\sqrt{|\|\vec{H}\|^2|}}.$$

Remark 2.2.8. It is clear from equation (2.6) that $\vec{H} \neq 0$ everywhere for a hyperquadric homothety of the mean curvature flow, so that ν can be defined and by the same equation: ν satisfies $\nabla^\perp \nu := \left(\nabla^{I^*(\mathbb{R}^n, \langle \cdot, \cdot \rangle)} \nu\right)^\perp = 0$ and $\vec{H}$ satisfies $\nabla^\perp \vec{H} = 0$.

Example 2.2.9 (The Hyperquadrics in Minkowski space). There are many works about homotheties of the MCF in $\mathbb{E}^n$, so that the most relevant new case in this work are the homotheties in the 4-dimensional Minkowski Space $\mathbb{R}^{1,4}$. The hyperquadrics in $\mathbb{R}^{1,3}$ are just a cone (for $k=0$), hyperboloids of one sheet (for $k>0$) or hyperboloids of two sheets (for $k<0$). If one writes an atlas for a hyperquadric in $\mathbb{R}^{1,n}$, one can see that $\mathcal{H}^{n-1}(k)$ is a spacelike for $k<0$ but that $T_p\mathcal{H}^{n-1}(k)$ has vectors with negative length for $k>0$ at any $p \in M$.

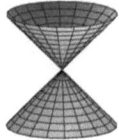 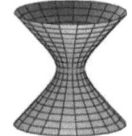 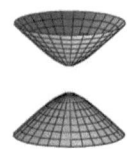

Figure 2.1: Light Cone   Figure 2.2: $\mathcal{H}^2(1)$ Figure 2.3: $\mathcal{H}^2(-1)$

Of course $\mathcal{H}^{n-1}(k)$ is a minimal submanifold of itself if $k<0$, so that the mean curvature flow of the natural immersion in $\mathbb{R}^{1,n}$ is just the homothety given in the last section. And we have long time existence for $k<0$. We saw that a minimal surface of $\mathcal{H}^{n-1}(k)$ for $k = \|F(0)\|^2 > 0$ converges pointwise to the origin of $\mathbb{R}^{1,n}$. Unfortunately, the metric of $\mathcal{H}^{n-1}(k)$ is nondegenerate for $k>0$, but if we consider the homothety $F(t) := c^{-1}(t)F(0)$ of the last section the whole hyperquadric converges in the Hausdorff norm to the light cone, as we prove now:

Let $Y = (Y_1,\ldots,Y_n)$ be a point in the light cone, so that $Y_1 = \pm\sqrt{Y_2^2+\ldots+Y_n^2}$.

- If $Y_2^2+\ldots+Y_n^2 \geq \|F(0)\|^2 - 2mt$, define

$$X_1 := \pm\sqrt{Y_2^2+\ldots+Y_n^2 - \|F(0)\|^2 + 2mt},$$

so that the point $X := (X_1, Y_2, \ldots, Y_n)$ is in $\mathcal{H}^{n-1}(\|F(0)\|^2 - 2mt)$. Then the Euclidean distance between X and Y is

$$d(X,Y) \leq \sqrt{Y_1^2 - X_1^2} = \sqrt{\|F(0)\|^2 - 2mt}.$$

- If $Y_2^2+\ldots+Y_n^2 < \|F(0)\|^2 - 2mt$ and $Y_2^2+\ldots+Y_n^2 \neq 0$, define

$$\alpha := \sqrt{\frac{\|F(0)\|^2 - 2mt}{Y_2^2+\ldots+Y_n^2}}$$

so that $\alpha > 1$ and the point $X := (0, \alpha Y_2, \ldots, \alpha Y_n)$ is in $\mathcal{H}^{n-1}(\|F(0)\|^2 - 2mt)$. Then the Euclidean distance between X and Y is

$$d(X,Y) \leq \sqrt{Y_1^2 + (1-\alpha)^2(Y_2^2 + \ldots + Y_n^2)}$$
$$\leq \sqrt{Y_1^2 - (1-\alpha^2)(Y_2^2 + \ldots + Y_n^2)} \leq \sqrt{\|F(0)\|^2 - 2mt}.$$

- If $Y_2^2 + \ldots + Y_n^2 = 0$ then $Y_1 = 0$. Define $X := (\sqrt{\|F(0)\|^2 - 2mt}, 0 \ldots, 0)$. This point is in the hyperquadric $\mathcal{H}^{n-1}(\|F(0)\|^2 - 2mt)$ and the Euclidean distance between X and Y is

$$d(X,Y) = \sqrt{\|F(0)\|^2 - 2mt}.$$

The three possibilities show that, for any Y in the light cone

$$\inf_{X \in \mathcal{H}^{n-1}(\|F(0)\|^2 - 2mt)} d(Y,X) \leq \sqrt{\|F(0)\|^2 - 2mt}$$

and it follows, denoting the light cone as $\mathcal{H}^{n-1}(0)$, that

$$\sup_{Y \in \mathcal{H}^{n-1}(0)} \inf_{X \in \mathcal{H}^{n-1}(\|F(0)\|^2 - 2mt)} d(Y,X) \leq \sqrt{\|F(0)\|^2 - 2mt},$$

but this goes to 0 as t goes to $\frac{\|F(0)\|^2}{2m}$.

Analogously we get that

$$\sup_{X \in \mathcal{H}^{n-1}(\|F(0)\|^2 - 2mt)} \inf_{Y \in \mathcal{H}^{n-1}(0)} d(Y,X) \leq \sqrt{\|F(0)\|^2 - 2mt} \to 0$$

as t goes to $\frac{\|F(0)\|^2}{2m}$.

This completes the example.

Chapter 3

Principal Normal Parallel in the Normal Bundle

The two types of homotheties (self-shrinkers and self-expanders) lead, after rescaling, to different equations $\vec{H} = -F^\perp$ or $\vec{H} = F^\perp$, as further explained. We restrict our attention in this chapter to the self-shrinkers of the MCF that have a special property (the principal normal is parallel in the normal bundle).

In chapter 2 we saw that the hyperquadric homotheties of the MCF have two properties:

- The mean curvature vector $\vec{H}$ is, at every point $x \in M$, not a null-vector.

- The principal normal $\nu := \dfrac{\vec{H}}{\sqrt{|\|\vec{H}\|^2|}}$ is parallel in the normal bundle, this means $\nabla^\perp \nu = 0$,

if one considers a complexification of the tangent and normal bundles, the second condition is equivalent[1] to the possibly imaginary vector field $\nu := \dfrac{\vec{H}}{\|\vec{H}\|}$ being parallel in the normal bundle.

In this chapter we prove that a compact spacelike self-shrinker cannot satisfy $\|\vec{H}\|^2 < m$ (in particular cannot be negative) for all $x \in M$ and we also prove that the two conditions above are enough, if the dimension of M is different from 1, to ensure that a self-shrinker is hyperquadric, as the following Theorem states:

---

[1] assuming $\|H\|^2 \neq 0 \,\forall x \in M$

Theorem 3.0.1. [2] *Let M be a closed smooth manifold and $F : M \to (\mathbb{R}^n, \langle \cdot, \cdot \rangle)$ be a smooth immersion, which is a spacelike*[3] *self-shrinker of the mean curvature flow, i.e. F satisfies,*

$$\vec{H} = -F^\perp. \tag{3.1}$$

Besides, assume $m := \dim(M) \neq 1$. Then the mean curvature vector $\vec{H}$ satisfies $\|\vec{H}\|^2(p) \neq 0$ for all $p \in M$ and the principal normal ν is parallel in the normal bundle ($\nabla^\perp \nu \equiv 0$) if, and only if, F is a minimal immersion in the hyperquadric $\mathcal{H}^{n-1}(m)$.

Instead of writing $F(\cdot, t)$ we shall write only F. Chapter 2 was the first part of this Theorem, we just need to prove now that $\|F\|^2 \equiv m$ for a spacelike self-shrinker of the MCF with principal normal parallel in the normal bundle. We prove this using the maximum principle on the function $\|F\|^2$. But to do this it is necessary first to deduce some equations related with this function.

3.1 Fundamental Equations

In this section we calculate several equations involving the Laplacian of some tensors like the second fundamental form, the mean curvature vector, the Riemannian curvature and others. These equations are of great use to find geometric information about the manifold. For this purpose we use three auxiliary tensors

$$P_{ij} := \langle \vec{H}, A_{ij} \rangle, \qquad Q_{ij} := \langle A_i^k, A_{kj} \rangle, \qquad S_{ijkl} := \langle A_{ij}, A_{kl} \rangle.$$

Using Gauß equation (eq. (1.8)) we write the Ricci curvature in terms of these tensors as

$$R_{ij} = g^{kl} R_{kilj} = g^{kl} \langle A_{lk}, A_{ji} \rangle - g^{kl} \langle A_{jk}, A_{li} \rangle = \langle \vec{H}, A_{ji} \rangle - \langle A_{jk}, A_i^k \rangle = P_{ij} - Q_{ij}. \tag{3.2}$$

In this notation the useful Simon's equation is written as

---

[2] As $\|\vec{H}\|^2 > 0$ this is a slight generalization of Smoczyk's result for spacelike minimal immersed manifolds of the hyperquadrics of positive squared norm.
[3] The norm of all the tangent directions need to have the same sign or we cannot say anything about the sign of the Laplacian of a function at a maximum point and thence not use the maximum principle

Proposition 3.1.1.

$$\nabla_k^\perp \nabla_l^\perp \vec{H} = \triangle^\perp A_{kl} + R_{kilj}A^{ij} - R_k^i A_{il} + Q_l^i A_{ik} - S_{kilj}A^{ij} \qquad (3.3)$$

Proof.

$$\begin{aligned}
\nabla_k^\perp \nabla_l^\perp \vec{H} &= \nabla_k^\perp \nabla_l^\perp g^{ij} A_{ij} = g^{ij} \nabla_k^\perp \nabla_l^\perp A_{lj} \\
&= g^{ij}\nabla_i^\perp \nabla_k^\perp A_{lj} - g^{ij} R_{lki}^t A_{tj}^\alpha \frac{\partial}{\partial y^\alpha} - g^{ij} R_{jki}^t A_{lt}^\alpha \frac{\partial}{\partial y^\alpha} + g^{ij} R_{\beta k i}^{\perp \alpha} A_{lj}^\beta \frac{\partial}{\partial y^\alpha} \\
&= \triangle^\perp A_{kl} - R_{tlki} A^{ti} - R_{tk} A_l^t + \langle A_{it}, A_l^i \rangle A_k^t - \langle A_{kt}, A_{li} \rangle A^{ti} \\
&= \triangle^\perp A_{kl} + R_{likj} A^{ij} - R_k^i A_{il} + Q_l^i A_{ik} - S_{kilj} A^{ij},
\end{aligned}$$

where we used the commutation formula (Lemma 1.3.5) from the first to the second line and the Codazzi equation (eq. (1.5)). $\square$

The equations that we calculated until now are general equations, in the sense that they are true for immersions that satisfy the necessary conditions in which they are stated. We shall, from now on, use the main assumption that we have a self-similar shrinking solution of the mean curvature flow. To be more specific:

Let $F : M \times [0, T) \to (\mathbb{R}^n, \langle \cdot, \cdot \rangle)$ be a solution of the mean curvature flow, this means a smooth family (depending on a parameter t in the interval $[0, T)$) of immersions that satisfies the equation

$$\frac{d}{dt} F = \vec{H}.$$

Moreover, suppose that this family is a homothety of the mean curvature flow, which means that there is a rescaling function $c : [0, T) \to (0, \infty)$, such that $c(t)F(t, M)$ is equal to $F(0, M)$ as a set, so that $c(t)F(t)$ only moves the points inside $F(M, 0)$, which means that $\frac{d}{dt}[c(t)F(t)] \in TM$ and, as in eq. (2.2), F satisfies

$$\vec{H} = -\frac{\dot{c}}{c} F^\perp.$$

From this equation it is possible to see that we can assume $\dot{c} \neq 0$ for all $t \in [0, T)$ because if $\dot{c}(t) = 0$ then $\vec{H}(t) = 0$ everywhere in M, which means that the flow has stopped. But the sign of $\dot{c}$ still plays an important role, if it is positive it means that:

- The rescaling function c is growing and thence the immersion F is decreasing,

- The mean curvature vector points in the opposite direction of $F^\perp$,

if $\dot{c}$ has negative sign, we have the contrary, as figure[4] 3.1 shows.

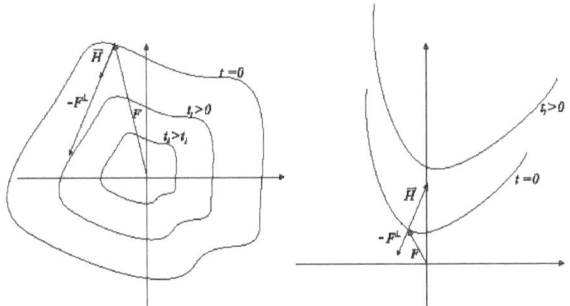

Figure 3.1: The two rescaling cases

For a fixed time $t \in [0,T)$ take $k > 0$, then consider another rescaled immersion: $\tilde{F} := kF$, which satisfies $\tilde{g} = k^2 g$, $\tilde{g}^{-1} = k^{-2} g^{-1}$ and $\vec{\tilde{H}} := \tilde{g}^{ij} \nabla_i \nabla_j \tilde{F} = k^{-1} \vec{H}$ so that this constant can be chosen as $k = \left|\frac{\dot{c}}{c}(t)\right|$ and one has $\vec{\tilde{H}} = -\tilde{F}^\perp$ or $\vec{\tilde{H}} = \tilde{F}^\perp$ depending on the directions of F and $\vec{H}$, as in the previous figure. We shall restrict ourselves to the self-shrinker case, this means we consider, from now on, a fixed $t \in [0,T)$ and the case

$$\vec{H} = -F^\perp.$$

On the other hand, from Huisken ([Hui90]), if an isometric immersion $G_0 : M \to (\mathbb{R}^n, \langle \cdot, \cdot \rangle)$ satisfies $\vec{H}_{G_0} = -G_0^\perp$ then the homothetic deformation $G : M \times [0, 1/2) \to (\mathbb{R}^n, \langle \cdot, \cdot \rangle)$ given by

$$G(x,t) := \sqrt{1-2t}\, G_0$$

---

[4] These two curves do not change homothetically with the MCF, these drawings just show that, in a general curve moving under some flow, $\vec{H}$ could point in the same direction of $F^\perp$ or in the opposite direction.

satisfies $G(x,0) = G_0$ and

$$\left(\frac{d}{dt}G(x,t)\right)^\perp = -(1-2t)^{-1/2}G_0^\perp = \vec{H}_{G(x,t)},$$

so that the deformation is (up to a tangential component) the mean curvature flow, but tangential components do not change the form of the immersed manifold, so that an immersion shrinks homothetically under the MCF if, and only if[5], it satisfies equation (3.4). This motivates the following definition:

Definition 3.1.2. Let $F : (M,g) \to (\mathbb{R}^n, \langle \cdot, \cdot \rangle)$ be an isometric immersion, i. e. $g := F^*\langle \cdot, \cdot \rangle$. We say that F is a *self-shrinker* (or a *self-similar shrinking solution*) of the MCF if

$$\vec{H} = -F^\perp. \tag{3.4}$$

From equation (3.4) it is clear that the projection of the position vector in the normal bundle is important for our calculations, and it can be easily written in terms of the tangential projection. This motivates us to define the following the 1-form:

$$\theta := \frac{1}{2}d\|F\|^2 = \langle F_i, F \rangle dx^i \tag{3.5}$$

such that $\theta^j F_j = \theta_i g^{ij} F_j$ is the pointwise projection of F in the tangent plane, with $\theta_i = \langle F_i, F \rangle$. Then we calculate:

$$\nabla_i \theta_j = \nabla_i \langle F_j, F \rangle = \langle A_{ij}, F \rangle + \langle F_i, F_j \rangle = \langle A_{ij}, F \rangle + g_{ij}.$$

We follow with

$$\nabla_i^\perp F^\perp = (\nabla_i(F - \theta^k F_k))^\perp = (F_i - \nabla_i \theta^k F_k - \theta^k A_{ik})^\perp = -\theta^k A_{ik} \tag{3.6}$$

and

$$\nabla_i^\perp \vec{H} = -\nabla_i^\perp F^\perp = \theta^k A_{ik}. \tag{3.7}$$

---

[5]Up to rescaling.

So that

$$\begin{aligned}
\nabla_i^\perp \nabla_j^\perp F^\perp &= -\nabla_i^\perp(\theta^k A_{jk}) = -(\nabla_i \theta^k A_{jk} + \theta^k \nabla_i A_{jk})^\perp \\
&= -(\nabla_i \theta^k A_{jk} + \theta^k \nabla_i A_{jk})^\perp = -(((\langle A_{il}, F\rangle + g_{il})g^{lk} A_{jk} + \theta^k \nabla_i A_{jk})^\perp \\
&= -A_{ij} - \langle A_i^k, F^\perp\rangle A_{jk} - \theta^k \nabla_k^\perp A_{ij}
\end{aligned}$$

where we used the Codazzi equation (Theorem 1.3.1) in the last step. From this follows that

$$\nabla_i^\perp \nabla_j^\perp \vec{H} = -\nabla_i^\perp \nabla_j^\perp F^\perp = A_{ij} - P_i^k A_{kj} + \theta^k \nabla_i^\perp A_{jk} \tag{3.8}$$

and

$$\triangle^\perp \vec{H} = g^{ij} \nabla_i^\perp \nabla_j^\perp \vec{H} = \vec{H} - P^{ik} A_{ik} + \theta^k \nabla_k^\perp \vec{H}. \tag{3.9}$$

Now we are able to calculate the Laplacian of the squared norm of the mean curvature vector.

$$\begin{aligned}
\triangle \|\vec{H}\|^2 &= g^{ij} \nabla_i \nabla_j \langle \vec{H}, \vec{H}\rangle = 2g^{ij} \nabla_i \langle \nabla_j \vec{H}, \vec{H}\rangle = 2g^{ij} \nabla_i \langle \nabla_j^\perp \vec{H}, \vec{H}\rangle \\
&= 2g^{ij}(\langle \nabla_i \nabla_j^\perp \vec{H}, \vec{H}\rangle + \langle \nabla_j^\perp \vec{H}, \nabla_i \vec{H}\rangle) \\
&= 2g^{ij}(\langle \nabla_i^\perp \nabla_j^\perp \vec{H}, \vec{H}\rangle + \langle \nabla_j^\perp \vec{H}, \nabla_i^\perp \vec{H}\rangle) \\
&= 2\langle \triangle^\perp \vec{H}, \vec{H}\rangle + 2\|\nabla^\perp \vec{H}\|^2 = 2\langle \vec{H} - P^{ik} A_{ik} + \theta^k \nabla_k^\perp \vec{H}, \vec{H}\rangle + 2\|\nabla^\perp \vec{H}\|^2 \\
\triangle \|\vec{H}\|^2 &= 2\|\vec{H}\|^2 - 2\|P\|^2 + 2\|\nabla^\perp \vec{H}\|^2 + \langle F^\top, \nabla\|\vec{H}\|^2\rangle, \tag{3.10}
\end{aligned}$$

because

$$\begin{aligned}
2\langle \theta^k \nabla_k^\perp \vec{H}, \vec{H}\rangle &= 2\langle F, F_l\rangle g^{lk} \langle \nabla_k \vec{H}, \vec{H}\rangle = \langle F, F_l\rangle g^{lk} \nabla_k \langle \vec{H}, \vec{H}\rangle \\
&= \langle \langle F, F_l\rangle g^{lu} F_u, \nabla_k \langle \vec{H}, \vec{H}\rangle g^{kt} F_t\rangle = \langle F^\top, \nabla\|\vec{H}\|^2\rangle.
\end{aligned}$$

For the squared norm of the second fundamental form, using Simon's equation (Theorem 3.1.1), one gets:

$$\begin{aligned}
2\langle A, (\nabla^\perp)^2 \vec{H}\rangle &= g^{tk}g^{sl}2\langle A_{ts}, \nabla_k^\perp \nabla_l^\perp \vec{H}\rangle \\
&= g^{tk}g^{sl}2\langle A_{ts}, \triangle^\perp A_{kl} + R_{kilj}A^{ij} - R_k^i A_{il} + Q_l^i A_{ik} - S_{kilj}A^{ij}\rangle \\
&= 2\langle A, \triangle^\perp A\rangle + 2\langle A^{kl}, R_{kilj}A^{ij}\rangle - 2\langle A^{kl}, R_k^i A_{il}\rangle \\
&\quad + 2\langle A^{kl}, Q_l^i A_{ik}\rangle - 2\langle A^{kl}, S_{kilj}A^{ij}\rangle \\
&= \triangle\|A\|^2 - 2\|\nabla^\perp A\|^2 + 2R_{kilj}S^{ijkl} \\
&\quad - 2R_{ij}Q^{ij} + 2\|Q\|^2 - 2S_{ikjl}S^{ijkl}.
\end{aligned}$$

On the other hand, using eq. (1.10) for the Ricci tensor on the normal bundle, we get

$$\begin{aligned}
\|R^\perp\|^2 &= \langle A_{jk}\otimes A_i^k - A_{ik}\otimes A_j^k, A_l^j \otimes A^{li} - A_l^i \otimes A^{jl}\rangle \\
&= \langle A_{jk}, A_l^j\rangle\langle A_i^k, A^{li}\rangle - \langle A_{ik}, A_l^j\rangle\langle A_j^k, A^{li}\rangle \\
&\quad - \langle A_{jk}, A_l^i\rangle\langle A_i^k, A^{lj}\rangle + \langle A_{ik}, A_l^i\rangle\langle A_j^k, A^{lj}\rangle \\
\|R^\perp\|^2 &= Q_{kl}Q^{kl} - S_{ikjl}S^{kjli} - S_{jkil}S^{kilj} + Q_{kl}Q^{kl} = 2\|Q\|^2 - 2S_{ikjl}S^{ijkl}.
\end{aligned}$$
(3.11)

So that, using these last two equations, we reach

$$\begin{aligned}
2\langle A, (\nabla^\perp)^2\vec{H}\rangle &= \triangle\|A\|^2 - 2\|\nabla^\perp A\|^2 + 2R_{kilj}S^{ijkl} - 2R_{ij}Q^{ij} + \|R^\perp\|^2 \\
&= \triangle\|A\|^2 - 2\|\nabla^\perp A\|^2 + 2\langle A_{ik}, A_{jl}\rangle S^{ijkl} \\
&\quad - 2\langle A_{il}, A_{jk}\rangle S^{ijkl} - 2(P_{ij} - Q_{ij})Q^{ij} + \|R^\perp\|^2 \\
&= \triangle\|A\|^2 - 2\|\nabla^\perp A\|^2 + 2S_{lkji}S^{kilj} \\
&\quad - 2S_{jkli}S^{kilj} - 2P_{ij}Q^{ij} + 2\|Q\|^2 + \|R^\perp\|^2 \\
2\langle A, (\nabla^\perp)^2\vec{H}\rangle &= \triangle\|A\|^2 - 2\|\nabla^\perp A\|^2 + 2\|S\|^2 - 2\langle P, Q\rangle + 2\|R^\perp\|^2, \quad (3.12)
\end{aligned}$$

where we used the Gauß equation (eq. (1.8)), equation (3.2) that relates the Ricci tensor to the tensors P and Q and eq. (3.11) in the last step.

On the other hand, we can calculate an equation for $\triangle\|A\|^2$ using Simon's equation (Theorem 3.1.1) in the following way:

First, with equations (3.8) and (3.1.1), we have

$$\begin{aligned}
\triangle^\perp A_{kl} =& \nabla_k^\perp \nabla_l^\perp \vec{H} - R_{kilj} A^{ij} + R_k^i A_{il} - Q_l^i A_{ik} + S_{kilj} A^{ij} \\
=& A_{kl} - P_k^i A_{il} + \theta^i \nabla_k^\perp A_{li} + R_k^i A_{il} - Q_l^i A_{ik} + (S_{kilj} - R_{kilj}) A^{ij} \\
\triangle^\perp A_{kl} =& A_{kl} - Q_k^i A_{il} - Q_l^i A_{ik} + \theta^t \nabla_k^\perp A_{lt} + (S_{kilj} - R_{kilj}) A^{ij},
\end{aligned} \qquad (3.13)$$

which implies

$$\begin{aligned}
\triangle \|A\|^2 =& g^{ij} \nabla_i \nabla_j \langle A_{kl}, A^{kl} \rangle = 2\langle \triangle^\perp A_{kl}, A^{kl} \rangle + 2 g^{ij} \langle \nabla_i^\perp A_{kl}, \nabla_j^\perp A^{kl} \rangle \\
=& 2\langle A_{kl} - Q_k^i A_{il} - Q_l^i A_{ik} + \theta^i \nabla_k^\perp A_{li} + (S_{kilj} - R_{kilj}) A^{ij}, A^{kl} \rangle + 2\|\nabla^\perp A\|^2 \\
=& 2\|A\|^2 - 4\|Q\|^2 + \langle F^\top, \nabla \|A\|^2 \rangle + 2(2 S_{kilj} - S_{klij}) S^{ijkl} + 2\|\nabla^\perp A\|^2 \\
\triangle \|A\|^2 =& 2\|A\|^2 - 2\|R^\perp\|^2 + \langle F^\top, \nabla \|A\|^2 \rangle - 2\|S\|^2 + 2\|\nabla^\perp A\|^2,
\end{aligned} \qquad (3.14)$$

where we used the Gauß equation (eq. (1.8)), equation (3.11) and

$$\begin{aligned}
2\langle \theta^i \nabla_k^\perp A_{li}, A^{kl} \rangle =& 2\langle F, F_t \rangle g^{ti} \langle \nabla_i A_{kl}, A^{kl} \rangle = \langle F, F_t \rangle g^{ti} \nabla_i \langle A_{kl}, A^{kl} \rangle \\
=& \langle \langle F, F_t \rangle g^{tu} F_u, \nabla_i \langle A_{kl}, A^{kl} \rangle g^{is} F_s \rangle = \langle F^\top, \nabla \|A\|^2 \rangle.
\end{aligned}$$

Theorem 3.1.3. *Let M be a closed smooth manifold and $F : M \to (\mathbb{R}^n, \langle \cdot, \cdot \rangle)$ be a spacelike self-shrinker of the mean curvature flow. Then it cannot hold $\|\vec{H}\|^2 < m := dim(M)$.*

Proof. If $\|\vec{H}\|^2 < m$ for all $x \in M$, then

$$\triangle \|F\|^2 = 2 g^{ij} \langle F_i, F_j \rangle + 2 \langle \triangle F, F \rangle = 2m - 2\|\vec{H}\|^2 > 0. \qquad (3.15)$$

Otherwise, as M is close, let $x \in M$ with $\|F(x)\|^2 = \max_{y \in M} \|F(y)\|^2$ and choose Riemannian normal coordinates for a neighborhood of x. Then $g_{ij} = \delta_{ij}$ and $\Gamma_{ij}^k = 0$ at x, thus

$$\triangle \|F\|^2(x) = \sum_i \frac{\partial^2}{\partial x^i \partial x^i} \|F\|^2(x).$$

Now let γ be the geodesic through x in the direction of $\frac{\partial}{\partial x^i}$. Then the restriction of $\|F\|^2$ over γ also has a maximum at x and

$$\frac{\partial^2}{\partial x^i \partial x^i}\|F\|^2(x) = \left(\|F\|^2\big|_\gamma\right)''(x) \leq 0$$

by the test of the second derivative for real functions, so that $\triangle \|F\|^2(x) \leq 0$, which contradicts (3.15). $\square$

Remark 3.1.4. In particular there are no spacelike self-shrinkers with $\|\vec{H}\|^2 < 0$ and no spacelike self-shrinkers if the index of $(\mathbb{R}^n, \langle \cdot, \cdot \rangle)$ is $n - m$.

3.2 The Compact Case

The special case that we consider now are the self-shrinkers of the MCF that satisfy the following conditions:

- The mean curvature vector is not a null vector

$$\|\vec{H}(x)\|^2 \neq 0, \text{ for all } x \in M.$$

- The principal normal $\nu := \frac{1}{\|\vec{H}\|}\vec{H}$ is parallel in the normal bundle

$$\nabla^\perp \nu \equiv 0,$$

where we write $\|\vec{H}\|$ to the complex function $\sqrt{\|\vec{H}\|^2} : M \to \mathbb{C}$, which is well defined (with $\sqrt{-1} = i$), because $\|\vec{H}\|^2 \neq 0$ everywhere and thus $\|\vec{H}\|^2$ does not change sign. Although Theorem 3.1.3 implies that $\|\vec{H}\|^2 \geq 0$ in the compact case, we also consider the possibility that $\|\vec{H}\|^2 \leq 0$ for the calculations in this chapter because we are going to need them in the non-compact case. A remark is that this definition is different from the motivation at the beginning of last section, but it is clear that the property of being parallel in the normal bundle is equivalent.

Note that the complex function $\|\vec{H}\|$ is a pure real or a pure imaginary all over M. So ν may not to be a real vector, but a vector field in the complexification of the pullback over M of $T\mathbb{R}^n$ ($F^{-1}T\mathbb{R}^n_\mathbb{C}$). Over this bundle, we use the complex linear extension ($\nabla^{\mathbb{C}}$) of the Levi-Civita connection in $(\mathbb{R}^n, \langle \cdot, \cdot \rangle)$ as connection. So that, for $Z \in \Gamma(F^{-1}T\mathbb{R}^n_{\mathbb{C}})$, with $Z = X + iY$ and $X, Y, V \in \Gamma(TM)$, we have

$$\nabla^{\mathbb{C}}_V Z = \nabla_V X + i\nabla_V Y.$$

We also need to extend the inner product $\langle, \rangle$ to the complex numbers and we do that linearly, too

$$\langle X_1 + iY_1, X_2 + iY_2 \rangle_{\mathbb{C}} = \langle X_1, X_2 \rangle - \langle Y_2, Y_2 \rangle + i(\langle X_1, Y_2 \rangle + \langle Y_1, X_2 \rangle).$$

To the metric in product bundles we shall use the metric g on the elements in TM and TM^* and the complexified inner product on the elements in $F^{-1}T\mathbb{R}^n_{\mathbb{C}}$.

Remark 3.2.1. It should be said that these new "metrics" are not semi-Riemannian metrics, because they deliver complex numbers.

Additionally, we still have to explain what we mean by the normal projection of a complex vector and that is, for $X \in \Gamma(F^*T\mathbb{R}^n)$, the real case projection $X^{\perp} = X - \langle X, F_i \rangle g^{ij} F_j$ and for pure imaginary vectors $(iX)^{\perp} = i(X^{\perp})$. We only consider projections of pure real or pure imaginary tensors.

Remark 3.2.2. An important remark is that the principal normal is pure real or pure imaginary, that the mean curvature vector is pure real and the second fundamental tensor is pure real and in the following there are no sums of a real and imaginary vector fields or sums of a real and imaginary functions. So that, because of the linearity of the extensions, we will have just whole real equations multiplied by i or just real equations and the metrics and connections just work as in the real case (maybe multiplied by i). Then we will not write the subscript indicating that the object is complex.

A parallel principal normal (in the normal bundle) can simplify some of the equations that we previously calculated because of its properties:

$$\nabla^{\perp}_k \vec{H} = \nabla^{\perp}_k (\|\vec{H}\|\nu) = \nabla_k \|\vec{H}\|\nu \tag{3.16}$$

and

$$\triangle^{\perp} \vec{H} = g^{ij}\nabla^{\perp}_i \nabla^{\perp}_j (\|\vec{H}\|\nu) = g^{ij}\nabla_i \nabla_j \|\vec{H}\|\nu = \triangle \|\vec{H}\|\nu. \tag{3.17}$$

From this, using equation (3.9), we calculate

$$P^{ij}A_{ij} = \vec{H} + \theta^k \nabla_k^\perp \vec{H} - \triangle^\perp \vec{H} = (\|\vec{H}\| + \theta^k \nabla_k \|\vec{H}\| - \triangle \|\vec{H}\|)\nu,$$

which means that $P^{ij}A_{ij}$ is a vector field in the same direction as ν (or $i\nu$, if ν is imaginary). Then follows the Lemma:

Lemma 3.2.3. *Let $F : M \to (\mathbb{R}^n, \langle \cdot, \cdot \rangle)$ be an immersion such that the principal normal is parallel in normal bundle, then*

1. $P^{ij}A_{ij} = \frac{\|P\|^2}{\|\vec{H}\|}\nu$ 2. $S_{ijkl}P^{ij}P^{kl} = \frac{\|P\|^4}{\|\vec{H}\|^2}$
3. $P_i^k A_{kj} = P_j^k A_{ki}$ 4. $S_{ikjl}P^{ij}P^{kl} = Q_{il}P_k^i P^{kl}$

Proof. 1. $P^{ij}A_{ij} = P^{ij}\langle \nu, A_{ij}\rangle \nu = \frac{P^{ij}}{\|\vec{H}\|}P_{ij}\nu = \frac{\|P\|^2}{\|\vec{H}\|}\nu,$

2. $S_{ijkl}P^{ij}P^{kl} = \langle A_{ij}P^{ij}, A_{kl}P^{kl}\rangle = \left\langle \frac{\|P\|^2}{\|\vec{H}\|}\nu, \frac{\|P\|^2}{\|\vec{H}\|}\nu \right\rangle = \frac{\|P\|^4}{\|\vec{H}\|^2}.$

3. To the third item we first need

$$\nabla_i^\perp \nabla_j^\perp \vec{H} - \nabla_j^\perp \nabla_i^\perp \vec{H} = (\nabla_i \nabla_j \|\vec{H}\| - \nabla_j \nabla_i \|\vec{H}\|)\nu$$
$$= \left(\frac{\partial^2 \|\vec{H}\|}{\partial x^i \partial x^j} - \frac{\partial \|\vec{H}\|}{\partial x^k}\Gamma_{ij}^k - \frac{\partial^2 \|\vec{H}\|}{\partial x^j \partial x^i} + \frac{\partial \|\vec{H}\|}{\partial x^k}\Gamma_{ji}^k\right)\nu = 0.$$

so that, using equation (3.8), we have

$$0 = \nabla_i^\perp \nabla_j^\perp \vec{H} - \nabla_j^\perp \nabla_i^\perp \vec{H}$$
$$0 = A_{ij} - P_i^k A_{kj} + \theta^k \nabla_i^\perp A_{jk} - A_{ji} + P_j^k A_{ki} - \theta^k \nabla_j^\perp A_{ik} = -P_i^k A_{kj} + P_j^k A_{ki},$$

thanks to the Codazzi equation (eq. (1.5)).

4. Finally, using item 3,

$$S_{ikjl}P^{ij}P^{kl} = \langle A_{ik}P^{ij}, A_{jl}P^{kl}\rangle$$
$$= \langle A_i^j P_k^i, A_{jl}P^{kl}\rangle = Q_{il}P_k^i P^{kl}.$$

□

Using these last equations we can prove:

Lemma 3.2.4. *Let $F : M \to (\mathbb{R}^n, \langle \cdot, \cdot \rangle)$ be a self-shrinker of the MCF such that the principal normal is parallel in normal bundle, then*

$$\frac{4}{\|\vec{H}\|^4} \left\langle \nabla^\perp \vec{H}, \nabla^\perp A_{ij} \right\rangle P^{ij} = \frac{2}{\|\vec{H}\|} \left\langle \nabla \|\vec{H}\|, \nabla \left(\frac{\|P\|^2}{\|\vec{H}\|^4} \right) \right\rangle + 4 \frac{\|P\|^2}{\|\vec{H}\|^6} \|\nabla \|\vec{H}\|\|^2. \tag{3.18}$$

Proof. We start calculating

$$\begin{aligned}
\left\langle \nabla^\perp \vec{H}, \nabla^\perp A_{ij} \right\rangle P^{ij} &= \nabla^k \|\vec{H}\| \left\langle \nu, \nabla_k^\perp A_{ij} \right\rangle P^{ij} \\
&= \nabla^k \|\vec{H}\| \nabla_k (\langle \nu, A_{ij} \rangle) P^{ij} \\
&= \nabla^k \|\vec{H}\| \nabla_k \left(\frac{P_{ij}}{\|\vec{H}\|} \right) P^{ij} \\
&= \nabla^k \|\vec{H}\| \left(\frac{\nabla_k P_{ij}}{\|\vec{H}\|} - \frac{\nabla_k \|\vec{H}\| P_{ij}}{\|\vec{H}\|^2} \right) P^{ij} \\
&= \frac{1}{2\|\vec{H}\|} \langle \nabla \|\vec{H}\|, \nabla |P|^2 \rangle - \frac{\|P\|^2}{\|\vec{H}\|^2} \|\nabla \|\vec{H}\|\|^2
\end{aligned}$$

and

$$\begin{aligned}
\left\langle \nabla \|\vec{H}\|, \nabla \left(\frac{\|P\|^2}{\|\vec{H}\|^4} \right) \right\rangle &= \left\langle \nabla \|\vec{H}\|, \frac{\nabla \|P\|^2}{\|\vec{H}\|^4} - \frac{4\|P\|^2 \|\vec{H}\|^3 \nabla \|\vec{H}\|}{\|\vec{H}\|^8} \right\rangle \\
&= \frac{\langle \nabla \|\vec{H}\|, \nabla \|P\|^2 \rangle}{\|\vec{H}\|^4} - 4 \frac{\|P\|^2}{\|\vec{H}\|^5} \|\nabla \|\vec{H}\|\|^2.
\end{aligned}$$

These two equations imply that

$$\begin{aligned}
\frac{4}{\|\vec{H}\|^4} \left\langle \nabla^\perp \vec{H}, \nabla^\perp A_{ij} \right\rangle P^{ij} &= \frac{2}{\|\vec{H}\|} \frac{\langle \nabla \|\vec{H}\|, \nabla |P|^2 \rangle}{\|\vec{H}\|^4} - 4 \frac{\|P\|^2}{\|\vec{H}\|^6} \|\nabla \|\vec{H}\|\|^2 \\
&= \frac{2}{\|\vec{H}\|} \left(\left\langle \nabla \|\vec{H}\|, \nabla \left(\frac{\|P\|^2}{\|\vec{H}\|^4} \right) \right\rangle + 4 \frac{\|P\|^2}{\|\vec{H}\|^5} \|\nabla \|\vec{H}\|\|^2 \right) \\
&\quad - 4 \frac{\|P\|^2}{\|\vec{H}\|^6} \|\nabla \|\vec{H}\|\|^2
\end{aligned} \tag{3.19}$$

$$\frac{4}{\|\vec{H}\|^4}\left\langle \nabla^\perp \vec{H}, \nabla^\perp A_{ij}\right\rangle P^{ij} = \frac{2}{\|\vec{H}\|}\left\langle \nabla\|\vec{H}\|, \nabla\left(\frac{\|P\|^2}{\|\vec{H}\|^4}\right)\right\rangle + 4\frac{\|P\|^2}{\|\vec{H}\|^6}\|\nabla\|\vec{H}\|\|^2. \tag{3.20}$$

□

We write some tensors with indices to avoid confusion when there is a sum of two tensors and it is not clear which term applies to each input of the tensor. We continue calculating,

$$\frac{2}{\|\vec{H}\|^4}\left\|\nabla_i\|\vec{H}\|\frac{P_{jk}}{\|\vec{H}\|} - \|\vec{H}\|\nabla_i\left(\frac{P_{jk}}{\|\vec{H}\|}\right)\right\|^2$$

$$=\frac{2}{\|\vec{H}\|^6}\|\nabla\|\vec{H}\|\|^2\|P\|^2 + \frac{2}{\|\vec{H}\|^2}\left\|\nabla_i\left(\frac{P_{jk}}{\|\vec{H}\|}\right)\right\|^2 - \frac{4}{\|\vec{H}\|^4}\nabla_i\|\vec{H}\|\nabla^i\left(\frac{P_{jk}}{\|\vec{H}\|}\right)P^{jk}$$

$$=\frac{2}{\|\vec{H}\|^6}\|\nabla\|\vec{H}\|\|^2\|P\|^2 + \frac{2}{\|\vec{H}\|^2}\left\|\nabla_i\left(\frac{P_{jk}}{\|\vec{H}\|}\right)\right\|^2$$
$$- \frac{2}{\|\vec{H}\|}\left\langle \nabla\|\vec{H}\|, \frac{\nabla\|P\|^2}{\|\vec{H}\|^4}\right\rangle + \frac{4}{\|\vec{H}\|^6}\|\nabla\|\vec{H}\|\|^2\|P\|^2$$

$$=\frac{6}{\|\vec{H}\|^6}\|\nabla\|\vec{H}\|\|^2\|P\|^2 + \frac{2}{\|\vec{H}\|^2}\left\|\nabla_i\left(\frac{P_{jk}}{\|\vec{H}\|}\right)\right\|^2$$
$$- \frac{2}{\|\vec{H}\|}\left\langle \nabla\|\vec{H}\|, \nabla\left(\frac{\|P\|^2}{\|\vec{H}\|^4}\right) + 4\frac{\|P\|^2}{\|\vec{H}\|^5}\nabla\|\vec{H}\|\right\rangle$$

$$=\frac{2}{\|\vec{H}\|^2}\left\|\nabla_i\left(\frac{P_{jk}}{\|\vec{H}\|}\right)\right\|^2 - 2\frac{\|P\|^2}{\|\vec{H}\|^6}\|\nabla\|\vec{H}\|\|^2 - \frac{2}{\|\vec{H}\|}\left\langle \nabla\|\vec{H}\|, \nabla\left(\frac{\|P\|^2}{\|\vec{H}\|^4}\right)\right\rangle$$

and

$$\frac{2}{\|\vec{H}\|^2}\left\|\nabla\left(\frac{P}{\|\vec{H}\|}\right)\right\|^2 = \frac{2}{\|\vec{H}\|^2}\left\|\frac{\nabla_i P_{jk}}{\|\vec{H}\|} - \frac{\nabla_i\|\vec{H}\|P_{jk}}{\|\vec{H}\|^2}\right\|^2$$

$$=\frac{2}{\|\vec{H}\|^2}\left(\left\|\frac{\nabla_i P_{jk}}{\|\vec{H}\|}\right\|^2 + \left\|\frac{\nabla_i\|\vec{H}\|P_{jk}}{\|\vec{H}\|^2}\right\|^2\right)$$

$$- \frac{4}{\|\vec{H}\|^5}\left\langle \nabla_i P_{jk}, \nabla_i\|\vec{H}\|P_{jk}\right\rangle$$

$$\frac{2}{\|\vec{H}\|^2}\left\|\nabla\left(\frac{P}{\|\vec{H}\|}\right)\right\|^2 = 2\frac{\|\nabla P\|^2}{\|\vec{H}\|^4} + 2\frac{\|P\|^2}{\|\vec{H}\|^6}\|\nabla\|\vec{H}\|\|^2 - \frac{2}{\|\vec{H}\|}\left\langle\frac{\nabla\|P\|^2}{\|\vec{H}\|^4},\nabla\|\vec{H}\|\right\rangle$$

$$= 2\frac{\|\nabla P\|^2}{\|\vec{H}\|^4} + 2\frac{\|P\|^2}{\|\vec{H}\|^6}\|\nabla\|\vec{H}\|\|^2$$

$$- \frac{2}{\|\vec{H}\|}\left\langle\nabla\left(\frac{\|P\|^2}{\|\vec{H}\|^4}\right) + 4\frac{\|P\|^2\nabla\|\vec{H}\|}{\|\vec{H}\|^5},\nabla\|\vec{H}\|\right\rangle$$

$$= 2\frac{\|\nabla P\|^2}{\|\vec{H}\|^4} - 6\frac{\|P\|^2}{\|\vec{H}\|^6}\|\nabla\|\vec{H}\|\|^2 - \frac{2}{\|\vec{H}\|}\left\langle\nabla\|\vec{H}\|,\nabla\left(\frac{\|P\|^2}{\|\vec{H}\|^4}\right)\right\rangle.$$

With this we get the equation:

$$\frac{2}{\|\vec{H}\|^4}\left\|\nabla_i\|\vec{H}\|\frac{P_{jk}}{\|\vec{H}\|} - \|\vec{H}\|\nabla_i\left(\frac{P_{jk}}{\|\vec{H}\|}\right)\right\|^2 = 2\frac{\|\nabla P\|^2}{\|\vec{H}\|^4} - 8\frac{\|P\|^2}{\|\vec{H}\|^6}\|\nabla\|\vec{H}\|\|^2 \quad (3.21)$$
$$- \frac{4}{\|\vec{H}\|}\left\langle\nabla\|\vec{H}\|,\nabla\left(\frac{\|P\|^2}{\|\vec{H}\|^4}\right)\right\rangle.$$

Lemma 3.2.5. *Let $F: M \times [0,T) \to (\mathbb{R}^n, \langle\cdot,\cdot\rangle)$ be a self-shrinker of the MCF such that $\|\vec{H}\|^2 \neq 0$ for all $x \in M$ and the principal normal is parallel in the normal bundle. Then*

$$\triangle\left(\frac{\|P\|^2}{\|\vec{H}\|^4}\right) = \frac{2}{\|\vec{H}\|^4}\left\|\nabla_i\|\vec{H}\|\frac{P_{jk}}{\|\vec{H}\|} - \|\vec{H}\|\nabla_i\left(\frac{P_{jk}}{\|\vec{H}\|}\right)\right\|^2 \quad (3.22)$$
$$+ \left\langle F^\top, \nabla\left(\frac{\|P\|^2}{\|\vec{H}\|^4}\right)\right\rangle - \frac{2}{\|\vec{H}\|}\left\langle\nabla\|\vec{H}\|,\nabla\left(\frac{\|P\|^2}{\|\vec{H}\|^4}\right)\right\rangle.$$

Proof. We begin using equations (3.9) and (3.13) to calculate

$$\triangle P_{ij} = \nabla^k\nabla_k\langle\vec{H},A_{ij}\rangle = \nabla^k(\langle\nabla_k^\perp\vec{H},A_{ij}\rangle + \langle\vec{H},\nabla_k^\perp A_{ij}\rangle)$$
$$= \langle\nabla^{k\perp}\nabla_k^\perp\vec{H},A_{ij}\rangle + 2\langle\nabla_k^\perp\vec{H},\nabla^{k\perp}A_{ij}\rangle + \langle\vec{H},\nabla^{k\perp}\nabla_k^\perp A_{ij}\rangle$$
$$= \langle\vec{H} - P^{kl}A_{kl} + \theta^k\nabla_k^\perp\vec{H},A_{ij}\rangle + 2\langle\nabla_k^\perp\vec{H},\nabla^{k\perp}A_{ij}\rangle$$
$$+ \langle\vec{H},A_{ij} - Q_i^kA_{kj} - Q_j^kA_{ki} + \theta^k\nabla_i^\perp A_{jk} + (S_{ikjl} - R_{ikjl})A^{kl}\rangle$$
$$= P_{ij} - P^{lk}S_{lkij} + \langle\theta^k\nabla_k^\perp\vec{H},A_{ij}\rangle + 2\langle\nabla_k^\perp\vec{H},\nabla^{k\perp}A_{ij}\rangle$$
$$+ P_{ij} - Q_i^kP_{kj} - Q_j^kP_{ki} + (S_{ikjl} - R_{ikjl})P^{kl} + \langle\vec{H},\theta^k\nabla_k^\perp A_{ij}\rangle$$

$$\begin{aligned}\triangle P_{ij} =& 2P_{ij} - P^{lk}S_{lkij} + 2\langle \nabla_k^\perp \vec{H}, \nabla^{k\perp}A_{ij}\rangle - Q_i^k P_{kj} - Q_j^k P_{ki} \\ &+ (S_{ikjl} - -S_{ijkl} + S_{ilkj})P^{kl} + \theta^k \nabla_k P_{ij} \\ =& 2P_{ij} + 2\langle \nabla^\perp \vec{H}, \nabla^\perp A_{ij}\rangle - Q_i^k P_{kj} - Q_j^k P_{ki} \\ &+ 2(S_{ikjl} - S_{ijkl})P^{kl} + \langle F^\top, \nabla P_{ij}\rangle,\end{aligned}$$

where we used the Gauß equation (eq. (1.8)) and, as in previous calculations, $\theta^k \nabla_k P_{ij} = \langle F^\top, \nabla P_{ij}\rangle$.

From this follows that

$$\begin{aligned}\triangle \|P\|^2 =& \triangle(P_{ij}P^{ij}) = 2\triangle P_{ij} P^{ij} + 2\langle \nabla P, \nabla P\rangle \\ =& 4\|P\|^2 + 4\langle \nabla^\perp \vec{H}, \nabla^\perp A_{ij}\rangle P^{ij} + 4(S_{ikjl} - S_{ijkl})P^{kl}P^{ij} \\ &- 4Q_i^k P_{kj} P^{ij} + 2\langle F^\top, \nabla P_{ij}\rangle P^{ij} + 2\|\nabla P\|^2 \\ =& 2\|\nabla P\|^2 + \langle F^\top, \nabla \|P\|^2\rangle + 4\langle \nabla^\perp \vec{H}, \nabla^\perp A_{ij}\rangle P^{ij} - 4\frac{\|P\|^4}{\|\vec{H}\|^2} + 4\|P\|^2,\end{aligned}$$

where we used Lemma 3.2.3 in the last step. On the other side

$$\begin{aligned}\triangle\left(\frac{\|P\|^2}{\|\vec{H}\|^4}\right) =& \nabla^i \nabla_i \left(\frac{\|P\|^2}{\|\vec{H}\|^4}\right) = \nabla^i \left(\frac{\nabla_i \|P\|^2}{\|\vec{H}\|^4} - 4\frac{\|P\|^2 \nabla_i \|\vec{H}\|}{\|\vec{H}\|^5}\right) \\ =& \frac{\triangle \|P\|^2}{\|\vec{H}\|^4} - 8\frac{\nabla_i \|P\|^2 \nabla^i \|\vec{H}\|}{\|\vec{H}\|^5} - 4\frac{\|P\|^2 \triangle \|\vec{H}\|}{\|\vec{H}\|^5} + 20\frac{\|P\|^2 \nabla_i \|\vec{H}\| \nabla^i \|\vec{H}\|}{\|\vec{H}\|^6} \\ =& \frac{\triangle \|P\|^2}{\|\vec{H}\|^4} - 8\frac{\nabla^i \|\vec{H}\|}{\|\vec{H}\|}\left(\frac{\nabla_i \|P\|^2}{\|\vec{H}\|^4} - 4\frac{\|P\|^2 \nabla_i \|\vec{H}\|}{\|\vec{H}\|^5}\right) \\ &- 2\frac{\|P\|^2}{\|\vec{H}\|^6}\left(2\triangle \|\vec{H}\| \|\vec{H}\| + 2\|\nabla \|\vec{H}\|\|^2\right) - 8\frac{\|P\|^2 \|\nabla\|\vec{H}\|\|^2}{\|\vec{H}\|^6},\end{aligned}$$

which implies that

$$\begin{aligned}\triangle\left(\frac{\|P\|^2}{\|\vec{H}\|^4}\right) =& \frac{\triangle \|P\|^2}{\|\vec{H}\|^4} - \frac{8}{\|\vec{H}\|}\left\langle \nabla \|\vec{H}\|, \nabla\left(\frac{\|P\|^2}{\|\vec{H}\|^4}\right)\right\rangle \\ &- 2\frac{\|P\|^2}{\|\vec{H}\|^6}\triangle \|\vec{H}\|^2 - 8\frac{\|P\|^2 \|\nabla \|\vec{H}\|\|^2}{\|\vec{H}\|^6}.\end{aligned} \quad (3.23)$$

Using the equations for $\triangle \|P\|^2$ and $\triangle \|\vec{H}\|^2$ (eq. (3.10)) we get

$$\triangle\left(\frac{\|P\|^2}{\|\vec{H}\|^4}\right) = \frac{2\|\nabla P\|^2 + \langle F^T, \nabla\|P\|^2\rangle + 4\langle \nabla^\perp \vec{H}, \nabla^\perp A_{ij}\rangle P^{ij}}{\|\vec{H}\|^4} - 8\frac{\|P\|^2 \|\nabla\|\vec{H}\|\|^2}{\|\vec{H}\|^6}$$
$$- \frac{8}{\|\vec{H}\|}\left\langle \nabla\|\vec{H}\|, \nabla\left(\frac{\|P\|^2}{\|\vec{H}\|^4}\right)\right\rangle - 2\frac{\|P\|^2}{\|\vec{H}\|^6}(2\|\nabla^\perp \vec{H}\|^2 + \langle F^T, \nabla\|\vec{H}\|^2\rangle)$$
$$= \left\langle F^T, \nabla\left(\frac{\|P\|^2}{\|\vec{H}\|^4}\right)\right\rangle + \frac{2\|\nabla P\|^2 + 4\langle \nabla^\perp \vec{H}, \nabla^\perp A_{ij}\rangle P^{ij}}{\|\vec{H}\|^4}$$
$$- \frac{8}{\|\vec{H}\|}\left\langle \nabla\|\vec{H}\|, \nabla\left(\frac{\|P\|^2}{\|\vec{H}\|^4}\right)\right\rangle - 12\frac{\|P\|^2}{\|\vec{H}\|^6}\|\nabla\|\vec{H}\|\|^2,$$

then we apply equations (3.21) and (3.20)

$$\triangle\left(\frac{\|P\|^2}{\|\vec{H}\|^4}\right) = \left\langle F^T, \nabla\left(\frac{\|P\|^2}{\|\vec{H}\|^4}\right)\right\rangle - \frac{2}{\|\vec{H}\|}\left\langle \nabla\|\vec{H}\|, \nabla\left(\frac{\|P\|^2}{\|\vec{H}\|^4}\right)\right\rangle$$
$$+ \frac{2}{\|\vec{H}\|^4}\left\|\nabla_i\|\vec{H}\|\frac{P_{jk}}{\|\vec{H}\|} - \|\vec{H}\|\nabla_i\left(\frac{P_{jk}}{\|\vec{H}\|}\right)\right\|^2,$$

where we used that $\nabla^\perp \vec{H} = \nabla\|\vec{H}\|\nu$. $\square$

In order to get some more specific (besides the equations that we already calculated) geometric information on the self-shrinkers of the mean curvature flow we need a further assumption: that M is closed, to effectively use the maximum principle. So, until the end of this chapter, let M be closed.

Proposition 3.2.6. *Let M be a closed smooth manifold and $F: M \to (\mathbb{R}^n, \langle\cdot,\cdot\rangle)$ be a smooth immersion, which is a spacelike self-shrinker of the mean curvature flow, i.e. F satisfies*

$$\vec{H} = -F^\perp.$$

Besides, assume that the mean curvature vector $\vec{H}$ satisfies $\|\vec{H}\|^2 \neq 0$ and the principal normal ν satisfies $\nabla^\perp \nu = 0$. Then

$$\left\|\nabla_i\|\vec{H}\|\frac{P_{jk}}{\|\vec{H}\|} - \|\vec{H}\|\nabla_i\left(\frac{P_{jk}}{\|\vec{H}\|}\right)\right\|^2 = 0. \tag{3.24}$$

Proof. First we note that although the function $\|\vec{H}\|$ is complex, the 3-tensor $\nabla_i\|\vec{H}\|\frac{P_{jk}}{\|\vec{H}\|} - \|\vec{H}\|\nabla_i\left(\frac{P_{jk}}{\|\vec{H}\|}\right)$ has only real coefficients, because $\|\vec{H}\|$ is either pure real or pure imaginary and $\|\vec{H}\|$ is in the numerators and denominators of the expression[6]. This implies that

$$\left\| \nabla_i \|\vec{H}\| \frac{P_{jk}}{\|\vec{H}\|} - \|\vec{H}\| \nabla_i \left(\frac{P_{jk}}{\|\vec{H}\|} \right) \right\|^2 \geq 0.$$

We first use the maximum principle on the function $\frac{\|P\|^2}{\|\vec{H}\|^4}$ to show that this function is constant:

From Lemma 3.2.5, we can write

$$\Delta \left(\frac{\|P\|^2}{\|\vec{H}\|^4} \right) \leq \left\langle F^T - \frac{2}{\|\vec{H}\|} \nabla \|\vec{H}\|, \nabla \left(\frac{\|P\|^2}{\|\vec{H}\|^4} \right) \right\rangle.$$

The function $u := \frac{\|P\|^2}{\|\vec{H}\|^4}$ has a maximum in M because M is closed. The point where the maximum is assumed is an interior point of some chart because M has no boundary. In this chart we write

$$g^{ij} \frac{\partial^2 u}{\partial x^i \partial x^j} - \left(g^{ij} \Gamma_{ij}^k + \langle F, F_l \rangle g^{lk} - \frac{2}{\|\vec{H}\|} \nabla_l \|\vec{H}\| g^{lk} \right) \frac{\partial u}{\partial x^k} \leq 0.$$

The strong elliptic maximum principle (Prop. 6.1.1) implies that u is constant in this coordinate neighborhood, and, from M being path connected, this can be extended to all charts, so that the coordinate independent function u satisfies:

$$\frac{\|P\|^2}{\|\vec{H}\|^4} = k \in \mathbb{R} \text{ with } k > 0,$$

then $\nabla \left(\frac{\|P\|^2}{\|\vec{H}\|^4} \right) = 0$ and $\Delta \left(\frac{\|P\|^2}{\|\vec{H}\|^4} \right) = 0$. Then theorem 3.2.5 implies that

$$\left\| \nabla_i \|\vec{H}\| \frac{P_{jk}}{\|\vec{H}\|} - \|\vec{H}\| \nabla_i \left(\frac{P_{jk}}{\|\vec{H}\|} \right) \right\|^2 = 0.$$

$\square$

We now rewrite the equality that we just proved in another way.

---

[6]this could also be seen rewriting the expression as $\frac{\nabla_i \|\vec{H}\|^2}{\|\vec{H}\|^2} P_{jk} - \nabla_i P_{jk}$.

First, we see that eq. (3.24) implies

$$\nabla_i\|\vec{H}\|\frac{P_{jk}}{\|\vec{H}\|} - \|\vec{H}\|\nabla_i\left(\frac{P_{jk}}{\|\vec{H}\|}\right) = 0, \tag{3.25}$$

as a tensor, because this is a pure covariant tensor over M and M is spacelike. Second, using the Codazzi equation (eq. (1.3.1)) and $\nabla^\perp \nu = 0$, we calculate

$$\nabla_i\left(\frac{P_{jk}}{\|\vec{H}\|}\right) = \nabla_i\langle \nu, A_{jk}\rangle = \langle \nu, \nabla_i^\perp A_{jk}\rangle$$
$$= \langle \nu, \nabla_j^\perp A_{ik}\rangle = \nabla_j\langle \nu, A_{ik}\rangle = \nabla_j\left(\frac{P_{ik}}{\|\vec{H}\|}\right).$$

Third, using equation (3.25), we write

$$\nabla_i\|\vec{H}\|\frac{P_{jk}}{\|\vec{H}\|} - \|\vec{H}\|\nabla_i\left(\frac{P_{jk}}{\|\vec{H}\|}\right) = \left(\nabla_i\|\vec{H}\|\frac{P_{jk}}{\|\vec{H}\|} - \nabla_j\|\vec{H}\|\frac{P_{ik}}{\|\vec{H}\|}\right)$$
$$+ \left(\nabla_j\|\vec{H}\|\frac{P_{ik}}{\|\vec{H}\|} - \|\vec{H}\|\nabla_i\left(\frac{P_{jk}}{\|\vec{H}\|}\right)\right)$$
$$= \left(\nabla_i\|\vec{H}\|\frac{P_{jk}}{\|\vec{H}\|} - \nabla_j\|\vec{H}\|\frac{P_{ik}}{\|\vec{H}\|}\right)$$
$$+ \left(\nabla_j\|\vec{H}\|\frac{P_{ik}}{\|\vec{H}\|} - \|\vec{H}\|\nabla_j\left(\frac{P_{ik}}{\|\vec{H}\|}\right)\right)$$
$$= \left(\nabla_i\|\vec{H}\|\frac{P_{jk}}{\|\vec{H}\|} - \nabla_j\|\vec{H}\|\frac{P_{ik}}{\|\vec{H}\|}\right),$$

which implies

$$0 = \left\|\nabla_i\|\vec{H}\|\frac{P_{jk}}{\|\vec{H}\|} - \|\vec{H}\|\nabla_i\left(\frac{P_{jk}}{\|\vec{H}\|}\right)\right\|^2 = \left\|\nabla_i\|\vec{H}\|\frac{P_{jk}}{\|\vec{H}\|} - \nabla_j\|\vec{H}\|\frac{P_{ik}}{\|\vec{H}\|}\right\|^2.$$

Now, expanding this norm we find

$$\|\nabla\|\vec{H}\|\|^2\frac{\|P\|^2}{\|\vec{H}\|^2} - \frac{\nabla_i\|\vec{H}\|}{\|\vec{H}\|^2}P_{jk}\nabla^j\|\vec{H}\|P^{ik} = 0$$
$$\|\nabla\|\vec{H}\|\|^2\|P\|^2 - \|\nabla_i\|\vec{H}\|P_k^i\|^2 = 0. \tag{3.26}$$

We are going to use this equation to show that the immersion F is hyperquadric, i. e. $\|F\|^2 = q \in \mathbb{R}$.

What remains to prove of Theorem 3.0.1 *Let M be a closed smooth manifold and $F: M \to (\mathbb{R}^n, \langle \cdot, \cdot \rangle)$ be an immersion, which is a spacelike self-shrinker of the mean curvature flow, i.e. F satisfies,*

$$\vec{H} = -F^\perp.$$

Besides, assume that the mean curvature vector $\vec{H}$ satisfies $\|\vec{H}\|^2 \neq 0$ and the principal normal ν satisfies $\nabla^\perp \nu = 0$. If $m := \dim(M) \neq 1$, then

$$\|F(x)\|^2 = m \, \forall x \in M.$$

Proof. We now calculate at a point $p \in M$ fixed. As the 2-tensor P is symmetric, it is also diagonalizable and has only real eigenvalues $\lambda_1, \ldots, \lambda_m$. Let $V_1, \ldots, V_m$ be an orthonormal basis of eigenvectors associated with $\lambda_1, \ldots, \lambda_m$. Then we write $\nabla \|\vec{H}\| = \Sigma_i \alpha_i V_i$, $\alpha_i \in \mathbb{C}$ so that by equation (3.26)

$$0 = \|P\|^2 \|\nabla\|\vec{H}\|\|^2 - \|P(\nabla\|\vec{H}\|)\|^2 = \sum_i \lambda_i^2 (\|\nabla\|\vec{H}\|\|^2 - \alpha_i^2), \qquad (3.27)$$

but $\lambda_i^2 \geq 0$ because $\lambda_i \in \mathbb{R}$, beyond this $\|\vec{H}\|$ is pure real or pure imaginary everywhere and all the α_i's have to agree with $\|\vec{H}\|$ about being real or imaginary, which implies that

$$\|\nabla\|\vec{H}\|\|^2 - \alpha_i^2 = \sum_{j \neq i} \alpha_j^2$$

is nonnegative for all $i \in \{1, \ldots, m\}$ if $\|\vec{H}\|$ is real and nonpositive for all $i \in \{1, \ldots, m\}$ if $\|\vec{H}\|$ is imaginary. This implies, with eq. (3.27), that

$$\lambda_i^2 (\|\nabla\|\vec{H}\|\|^2 - \alpha_i^2) = 0 \ \forall i \in \{1, \ldots, m\}.$$

As $tr(P) = P_{ij} g^{ij} = \|\vec{H}\|^2 \neq 0$, it follows that $P \neq 0$ and there is at least a $j \in \{1, \ldots, n\}$ such that $\lambda_j \neq 0$ and the last equation implies that

$$0 = \|\nabla\|\vec{H}\|\|^2 - \alpha_j^2 = \sum_i \alpha_i^2 - \alpha_j^2 = \sum_{i \neq j} \alpha_i^2 \implies \alpha_i = 0 \ \forall i \neq j,$$

because the α_i's are all real or all imaginary. From this follows that $\|\nabla\|\vec{H}\|\|^2 = \alpha_j^2$ and $\nabla\|\vec{H}\| = \alpha_j V_j$.

Now let us assume, by contradiction, that there is an $x \in M$ such that $\nabla\|\vec{H}\| \neq 0$ at this point.

Then $\alpha_j \neq 0$ and for all $i \neq j$

$$0 = \lambda_i^2(\|\nabla\|\vec{H}\|\|^2 - \alpha_i^2) = \lambda_i^2 \alpha_j^2 \implies \lambda_i = 0,$$

so that P_{ij} has only one nonzero eigenvalue and the associated eigenvector is $\nabla\|\vec{H}\|/\|\nabla\|\vec{H}\|\|$.

At this point we have

$$\|P\|^2 = \lambda_j^2 = (\operatorname{tr} P)^2 = \|\vec{H}\|^4 \implies \frac{\|P\|^2}{\|\vec{H}\|^4} = 1,$$

but we have already shown that this quotient is constant, so that the equation $\frac{\|P\|^2}{\|\vec{H}\|^4} = 1$ holds not only at this point but everywhere in M.

Then, using $\|P\|^2 = \|\vec{H}\|^4$, with equation (3.10) we calculate

$$2\|\nabla\|\vec{H}\|\|^2 + 2\|\vec{H}\|\triangle\|\vec{H}\| = \triangle\|\vec{H}\|^2 = 2\|\vec{H}\|^2 - 2\|P\|^2 + 2\|\nabla^\perp \vec{H}\|^2 + \langle F^\perp, \nabla\|\vec{H}\|^2\rangle$$
$$2\|\vec{H}\|\triangle\|\vec{H}\| = 2\|\vec{H}\|^2 - 2\|\vec{H}\|^4 + 2\|\vec{H}\|\langle F^\top, \nabla\|\vec{H}\|\rangle$$

and it follows

$$\triangle\|\vec{H}\| = \|\vec{H}\| - \|\vec{H}\|^3 + \langle F^\top, \nabla\|\vec{H}\|\rangle. \tag{3.28}$$

We integrate both sides of this equation. First integrate the terms of it separately taking advantage of the fact that M is closed:

$$\int_M \triangle\|\vec{H}\| = 0,$$

because of the Divergence Theorem[7], using that $\triangle u = \operatorname{div} \circ \operatorname{grad} u$, and

$$\int_M \langle F^\top, \nabla\|\vec{H}\|\rangle = \int_M \langle F, F_l\rangle g^{lk} \nabla_k \|\vec{H}\| = -\int_M \nabla_k \langle F, F_l\rangle g^{lk} \|\vec{H}\|$$
$$= -\int_M \langle F_k, F_l\rangle g^{lk} \|\vec{H}\| + \langle F, A_{kl}\rangle g^{lk} \|\vec{H}\| = -m\int_M \|\vec{H}\| + \int_M \|\vec{H}\|^3,$$

---

[7] $\int_M \operatorname{div}(X) = \int_{\partial M} \langle X, \vec{n}\rangle$ with $\vec{n}$ the exterior normal vector on ∂M.

such that

$$0 = \int_M \triangle \|\vec{H}\| = \int_M \|\vec{H}\| - \|\vec{H}\|^3 + \langle F^\perp, \nabla|\vec{H}|\rangle = (1-m)\int_M \|\vec{H}\|,$$

which is impossible for $m \neq 1$.

From this contradiction we know that $\nabla\|\vec{H}\| = 0$ everywhere in M and it follows that $\nabla^\perp \vec{H} = \nabla\|\vec{H}\|\nu = 0$ and, more importantly, that the norm of $\vec{H}$ is constant.

On the other hand

$$\triangle \|F\|^2 = 2\langle F_i, F_j\rangle g^{ij} + 2\langle \triangle F, F\rangle = 2g_{ij}g^{ij} + 2\langle \vec{H}, F\rangle = 2m - 2\|H\|^2.$$

If the constant $2m - 2\|H\|^2$ is other than zero (for example > 0) it would lead to a contradiction with the second derivative's test (at a local maximum of $\|F\|^2$ holds $\triangle\|F\|^2 \leq 0$), so that $\triangle\|F\|^2 = 0$ everywhere in M.

Again using the maximum principle (Prop. 6.1.1), we find that $\|F\|^2$ is constant. We can calculate this norm in that we see $\theta = d\|F\|^2 = \nabla\|F\|^2 = 0$, which implies that $F \in \Gamma(TM^\perp)$, so that $\vec{H} = -F$ and replacing $\|F\|^2 = \|\vec{H}\|^2$ and $\triangle\|F\|^2 \leq 0$ in eq. (3.2) we get $\|\vec{H}\|^2 = m$. $\square$

Note that the condition $\dim(M) \neq 1$ is optimal, because the result does not hold for the curve shortening flow[8], because the Abresch & Langer curves are not contained in a circle.

The following picture is an example of a closed geodesic (an ellipse) in the 1-leaf hyperboloid in $\mathbb{R}^{1,3}$. It is a minimal submanifold of $\mathcal{H}^2(1)$ and thus a spacelike self-shrinker of the MCF.

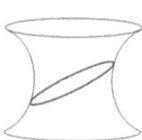

---
[8]The MCF for plane curves.

Chapter 4

The Non-Compact Case

We now consider non-compact self-shrinkers. The maximum principle that we used in the compact case cannot be used here. A tool in this case is to integrate over the whole manifold with respect to some function (a backwards heat kernel) and use partial integration to get geometric information from the equations. For that we are considering that $F(M)$ is unbounded, so that M itself is noncompact. By unbounded we mean:

Definition 4.0.1. Let us consider in $\mathbb{R}^n$ the usual topology, i. e. the topology of the euclidean open balls. A set $B \subset \mathbb{R}^n$ is *unbounded* if there is no compact set C, with $B \subset C$.

Remark 4.0.2. If $F(M)$ is unbounded, then, in particular, it is not contained in any euclidean sphere and thence $\|F\|_\mathbb{E}^2$ is also unbounded, i. e. for every $k \in \mathbb{R}$ there is a $x \in M$ with $\|F\|_\mathbb{E}^2(x) \geq k$.

In the pseudo-euclidean case there are minimal submanifolds of the hyperquadrics (hyperboloids), which are noncompact and are homotheties of the mean curvature flow with principal normal parallel in the normal bundle. But all of these hyperquadrics are asymptotic to the light cone and, in particular, have the norm $\|F\|^2$ bounded (constant). But these hyperquadrics do not satisfy the contitions that we need to integrate, so they do not appear in our results.

In the compact case (chapter 3) we proved that $\|\vec{H}\|^2 < m$ implies that F is not a self-shrinker of the MCF; in the non-compact case a similar result holds if M is stochastic complete[1] and $\sup_M \|F\|^2 < +\infty$.

---
[1]See definition 6.1.2 in the section Maximum principles in appendix.

Theorem 4.0.3. *The mean curvature vector of a stochastic complete, spacelike, self-shrinker of the mean curvature flow $F : M \to (\mathbb{R}^n, \langle \cdot, \cdot \rangle)$ cannot satisfy, for all $p \in M$,*
$$\|\vec{H}\|^2 < m - \epsilon,$$
for some $\epsilon > 0$ if $\sup_M \|F\|^2 < +\infty$.

Proof. If there is an $\epsilon > 0$ such that $\|\vec{H}\|^2 < m - \epsilon$ for all $x \in M$, then
$$\triangle \|F\|^2 = 2g^{ij}\langle F_i, F_j \rangle + 2\langle \triangle F, F \rangle = 2m - 2\|\vec{H}\|^2 > 2\epsilon,$$
but using the weak Omori-Yau maximum principle (Prop. 6.1.3, item 4) there is a sequence $\{x_k\}$ of points in M with the property
$$\triangle \|F\|^2(x_k) \leq \frac{1}{k},$$
which contradicts $\triangle \|F\|^2(x) > 2\epsilon$ for all $x \in M$. □

Remark 4.0.4. In particular, there are no stochastic complete, spacelike self-shrinkers of the mean curvature flow with $\sup_M \|F\|^2 < +\infty$ and $\|\vec{H}\|^2 \leq 0$.

Let us take a closer look to the pseudo-euclidean case.

Definition 4.0.5. Let $(\mathbb{R}^n, \langle \cdot, \cdot \rangle)$ be an inner product space and $\{e_1, \ldots, e_n\}$ an orthonormal basis such that $\langle e_\alpha, e_\alpha \rangle = -1$ for $\alpha \in \{1, \ldots, q\}$ and $\langle e_\alpha, e_\alpha \rangle = 1$ for $\alpha \in \{q+1, \ldots, n\}$, which we denote $\mathbb{R}^{q,n}$. For a vector $X \in \mathbb{R}^{q,n}$ we define his negative (X_-) and positive (X_+) projections as
$$X_- := -\sum_{\alpha=1}^{q} \langle X, e_\alpha \rangle e_\alpha$$
and
$$X_+ := \sum_{\alpha=q+1}^{n} \langle X, e_\alpha \rangle e_\alpha.$$

Remark 4.0.6. One sees immediately that $\|X\|_{\mathbb{E}}^2 = \|X_+\|^2 - \|X_-\|^2$ and $\|X\|^2 = \|X_+\|^2 + \|X_-\|^2$.

Remark 4.0.7. It also holds that $X = X_+ + X_-$ and $\|X\|^2 = 0$ exactly when $\|X_+\|^2 = -\|X_-\|^2$. So that $X \in (\mathbb{R}^n, \langle \cdot, \cdot \rangle)$, $X \neq 0$, is in the light cone if, and

only if,
$$-\frac{\|X_-\|^2}{\|X_+\|^2} = 1.$$

These projections can be extend for tensors. Let $A \in \Gamma(F^*T\mathbb{R}^{q,n} \otimes T^*M \otimes \ldots \otimes T^*M \otimes TM \otimes \ldots \otimes TM)$. Identifying $F_p^*T\mathbb{R}^{q,n}$ with $\mathbb{R}^{q,n}$, for $p \in M$, considering $\{e_1, \ldots, e_n\}$ an orthonormal basis as in the last definition to $T_{F(p)}\mathbb{R}^{q,n}$ and taking the inner product just in the component of the tensor that lies in $F^*T\mathbb{R}^{q,n}$, i. e.

$$A_- := -\sum_{\alpha=1}^{q} \langle A, e_\alpha \rangle \otimes e_\alpha$$

and

$$A_+ := \sum_{\alpha=q+1}^{n} \langle A, e_\alpha \rangle \otimes e_\alpha.$$

Remark 4.0.8. Note that $\mathbb{R}^{q,n}$ is the direct sum of two vector subspaces

$$V_1 := \{X \in \mathbb{R}^{q,n} | X = (X^1, \ldots, X^q, 0, \ldots, 0)\}$$

and

$$V_2 := \{X \in \mathbb{R}^{q,n} | X = (0, \ldots, 0, X^{q+1}, \ldots, X^n)\},$$

which are orthogonal.

In V_2 the inner product induced by $\langle \cdot, \cdot \rangle$ is positive definite and one can use the Cauchy-Schwarz inequality, for $V, W \in V_2$

$$|\langle V, W \rangle| \leq \|V\| \|W\|;$$

but in V_1 the inner product induced by $\langle \cdot, \cdot \rangle$ is negative definite and one can use the Cauchy-Schwarz inequality for $-\langle \cdot, \cdot \rangle$, so that $V, W \in V_1$

$$|-\langle V, W \rangle| \leq \sqrt{(-\|V\|^2)}\sqrt{(-\|W\|^2)}.$$

Let us now see more precisely how the hyperquadrics are asymptotic to the light-cone. Let $X \in \mathcal{H}^{n-1}(c)$, $c \neq 0$, with $\|X\|_{\mathbb{E}}^2 = k$, then:

$$\|X_+\|^2 - \|X_-\|^2 = k$$
$$\|X_+\|^2 + \|X_-\|^2 = c$$

and it follows
$$k+2\|X_-\|^2 = c, \qquad 2\|X_+\|^2 - k = c$$
so that
$$\|X_-\|^2 = \frac{c-k}{2}, \qquad \|X_+\|^2 = \frac{c+k}{2}$$
and, if $\|X_+\|^2 \neq 0$ (this is the case if $k > -c$),
$$-\frac{\|X_-\|^2}{\|X_+\|^2} = \frac{k-c}{k+c}.$$

But the hyperquadrics are unbounded, this means that for every $\epsilon > 0$ one can choose $k_\epsilon \in \mathbb{R}$ large enough, so that
$$1 - \epsilon < -\frac{\|X_-\|^2}{\|X_+\|^2} < 1 + \epsilon$$
for all $X \in \mathcal{H}^{n-1}(c)$ with $\|X\|_\mathbb{E}^2 \geq k_\epsilon$.

We consider the case, somewhat opposed to the hyperquadrics above considered, where $F(M)$ goes away from the light cone. More precisely:

Definition 4.0.9. Let M be a smooth manifold and $F : M \to \mathbb{R}^{q,n}$ be an immersion with $F(M)$ unbounded. We say that F (or $F(M)$) is *mainly positive* if there is an $\epsilon > 0$ and $k \in \mathbb{R}$, such that $\forall x \in M$
$$\|F(x)\|_\mathbb{E}^2 \geq k \implies -\frac{\|F(x)_-\|^2}{\|F(x)_+\|^2} \leq 1 - \epsilon.$$

And we say that F (or $F(M)$) is *mainly negative* if there is an $\epsilon > 0$ and $k \in \mathbb{R}$, such that $\forall x \in M$
$$\|F(x)\|_\mathbb{E}^2 \geq k \implies -\frac{\|F(x)_+\|^2}{\|F(x)_-\|^2} \leq 1 - \epsilon.$$

This means that there is an (Euclidean) angle θ with $\tan\left(\frac{\pi}{4} - \theta\right) < 1 - \epsilon$ between $F(x)$ and the light cone for any $x \in M$ such that $F(x)$ lies outside some big euclidean sphere (or $\tan\left(\frac{\pi}{4} - \theta\right) > 1 + \epsilon$ in the mainly negative case), as in figure 4.1.

Now we consider the behavior of $\|F(x)\|^2$ for x in M outside these big Euclidean spheres.

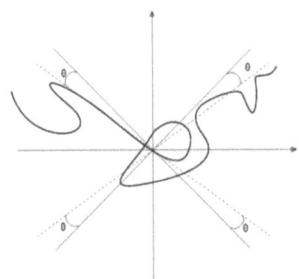

Figure 4.1: Mainly Positive Case

Lemma 4.0.10. *If $F(M)$ is mainly positive and unbounded, then $\|F\|^2 \geq \frac{\epsilon}{2}\|F(x)\|_{\mathbb{E}}^2$ and $\|F\|^2$ is unbounded.*

Proof. If $F(M)$ is mainly positive it holds, for $x \in M$ such that $\|F(x)\|_{\mathbb{E}}^2 > k$, with k and ϵ as in def. 4.0.9:

$$\|F\|^2(x) = \|F(x)_-\|^2 + \|F(x)_+\|^2 \geq \epsilon\|F(x)_+\|^2$$
$$\geq \frac{\epsilon}{2}\left(\|F(x)_+\|^2 - \|F(x)_-\|^2\right) = \frac{\epsilon}{2}\|F(x)\|_{\mathbb{E}}^2,$$

because $\|F(x)_+\|^2 \geq -\|F(x)_-\|^2$. This means that $\|F\|^2 \geq \frac{\epsilon}{2}\|F(x)\|_{\mathbb{E}}^2$, but $F(M)$ unbounded, so that $\|F\|^2$ is unbounded. $\square$

Lemma 4.0.11. *If $F(M)$ is mainly negative and unbounded, then $-\|F\|^2 \geq \frac{\epsilon}{2}\|F(x)\|_{\mathbb{E}}^2$ and $\|F\|^2$ is unbounded.*

Proof. If $F(M)$ is mainly negative it holds, for $x \in M$ such that $\|F(x)\|_{\mathbb{E}}^2 > k$, with k and ϵ as in def. 4.0.9,

$$-\|F\|^2 = -\|F(x)_-\|^2 - \|F(x)_+\|^2 \geq -\epsilon\|F(x)_-\|^2$$
$$\geq -\frac{\epsilon}{2}\left(\|F(x)_-\|^2 - \|F(x)_+\|^2\right) = \frac{\epsilon}{2}\|F(x)\|_{\mathbb{E}}^2,$$

because $\|F(x)_+\|^2 \leq -\|F(x)_-\|^2$. This means that $-\|F\|^2 \geq \frac{\epsilon}{2}\|F(x)\|_{\mathbb{E}}^2$, but $F(M)$ unbounded, so that $\|F\|^2$ is unbounded. $\square$

Remark 4.0.12. *If F is a spacelike self-shrinker such that $F(M)$ is mainly negative and unbounded then for $x \in M$ such that $\|F(x)\|_{\mathbb{E}}^2 > k$, with k as in* def. 4.0.9, it holds that

$$0 > \|F(x)\|^2 = \|F^\perp(x)\|^2 + \|F^\top(x)\|^2 \geq \|\vec{H}(x)\|^2,$$

but if M is stochastic complete, then Theorem 4.0.3 implies that F cannot be a self-shrinker of the MCF with $\|\vec{H}\|^2(p) \neq 0$ for all $p \in M$.

Definition 4.0.13. Let M be a smooth manifold and $f, g : M \to \mathbb{C}$ be continuous functions. We say that f *grows polynomially* with respect to g if there is some polynomial $\mathfrak{P} : \mathbb{R} \to \mathbb{R}$ such that

$$|f(x)| \leq \mathfrak{P}(|g(x)|) \qquad \forall\, x \in N,$$

where $|\cdot|$ is the norm of the complex number, $|a| = \sqrt{a\bar{a}}$.

In order to integrate we will still need to assume that the immersion F is "nice":

Definition 4.0.14. Let $F : M \to \mathbb{R}^{q,n}$ be a spacelike isometric immersion. We say that F has *bounded geometry* if:

1. There are $c_k, d_k \in \mathbb{R}$ for every $k \in \mathbb{N} \cup \{0\}$ such that[2].

$$\|(\nabla)^k A_+\|^2 \leq c_k,$$
$$-\|(\nabla)^k A_-\|^2 \leq d_k.$$

2. The function $\frac{1}{\|\vec{H}\|}$ grows polynomially with respect to $\|F\|^2$.

3. The growth of volume of geodesic balls and their boundaries is polynomial with respect to the radius. This means that there are polynomials $\mathfrak{P}, \mathfrak{Q} : \mathbb{R} \to \mathbb{R}$ such that, for any $p \in M$ and $R > 0$.

$$\mathrm{vol}(B_p(R)) \leq \mathfrak{P}(R) \qquad \mathrm{vol}(\partial B_p(R)) \leq \mathfrak{Q}(R),$$

---

[2] In [Smo05] polynomial growth (with respect to $\|F\|^2$) of $\|(\nabla)^k A\|^2$ is enough to guarantee integrability and some bounded growth on $\|A\|^2$ over some integral curves, but in our case the polynomial growth would guarantee the integrability but not the bounded growth on $\|A\|^2$ over some integral curves that are needed below.

where $B_p(R)$ is the geodesic ball of radius R and center p and $\partial B_p(R)$ its boundary. The volume $\text{vol}(\partial B_p(R))$ of the boundary of the geodesic ball is the $m-1$ dimensional volume.

4. F is *inverse Lipschitz* with respect to the euclidean norm in $\mathbb{R}^n$. This means, by definition, that there is a constant $k \in \mathbb{R}$ such that

$$\|F(x) - F(y)\|_\mathbb{E} \geq k d(x,y) \qquad \forall x,y \in M,$$

where $d(x,y)$ is the distance induced by the metric g on M.

To integrate we need some control on geodesic balls of M, which we get in the following Propositions.

Lemma 4.0.15. *Let $F : M \to \mathbb{R}^{q,n}$ be an inverse Lipschitz immersion with respect to the euclidean norm in $\mathbb{R}^n$, $\Omega_R := \{X \in F(M) \subset \mathbb{R}^{q,n} : \|X\|_\mathbb{E} < R\}$ and $p \in M$ be a fixed point such that $F(p) \in \Omega_R$. Then there is an (open) geodesic ball $B_{R'}(p)$ of radius $R' = 2R/k$, where k is the constant in the inverse Lipschitz condition, and center p such that $F^{-1}(\Omega_R) \subset B_{R'}(p)$.*

Proof. For any $X \in \Omega$ let $x \in F^{-1}(X)$, then

$$2R \geq \|F(p)\|_\mathbb{E} + \|F(x)\|_\mathbb{E} \geq \|F(p) - F(x)\|_\mathbb{E} \geq k d(p,x),$$

which means that $x \in B_{R'}(p)$. $\square$

Corollary 4.0.16. *Let Ω, R, $p \in F^{-1}(\Omega)$ and R' be as in the last Lemma and $y \in M$. Then*

$$d(p,y) > R' \quad \Rightarrow \quad y \notin \Omega,$$

this means $\|F(y)\|_\mathbb{E} > R$.

We continue by proving results for mainly positive immersions (the mainly negative ones will be discarded further below).

Remark 4.0.17. From Lemma 4.0.10: $\|F(x)\|_\mathbb{E}^2 \leq \frac{2}{\epsilon}\|F(x)\|^2$.

In the next Lemma we get a polynomial control of the radius of big geodesic balls in terms of $\|F\|^2$.

Lemma 4.0.18. *Let $F : M \to \mathbb{R}^{q,n}$ be a mainly positive, inverse Lipschitz immersion and $p \in M$. Then there is $R \in \mathbb{R}$ and $k_1, k_2 \in \mathbb{R}$ such that $x \notin B_p(R)$ implies $d(p,x) \leq k_1 \|F(x)\| + k_2$.*

Proof. Let $(1 >)\epsilon > 0$ and k be such as in the definition of mainly positive (def. 4.0.9). Now we consider $\Omega := \{X \in F(M) | \|X\|_\mathbb{E} < k\}$ and let R be big enough so that $F^{-1}(\Omega) \subset B_p(R)$ (by Lemma 4.0.15 assuming, without loss of generality, that $p \in F^{-1}(\Omega)$). Let $x \in M$, be such that $x \notin B_p(R)$, then $\|F(x)\|_\mathbb{E} \geq k$ and, by Rem. 4.0.17, it holds

$$\|F(x)\|_\mathbb{E}^2 \leq \frac{2}{\epsilon}\|F(x)\|^2.$$

But $F : M \to \mathbb{R}^{q,n}$ is inverse Lipschitz, so that there is $k_0 \in \mathbb{R}$ with

$$k_0 d(p,x) \leq \|F(p) - F(x)\|_\mathbb{E} \leq \|F(p)\|_\mathbb{E} + \|F(x)\|_\mathbb{E} \leq \|F(p)\|_\mathbb{E} + \sqrt{\frac{2}{\epsilon}}\|F(x)\|,$$

which completes the proof taking $k_1 = \frac{1}{k_0}\sqrt{\frac{2}{\epsilon}}$ and $k_2 = \|F(p)\|_\mathbb{E}/k_0$. $\square$

Remark 4.0.19. From Lemma 4.0.18 it follows that the distance to a fixed point $p \in M$ grows polynomially with respect to $\|F\|^2$ and thence the radius of a big ball has to be smaller than some polynomial of the infimum of $\|F\|^2$ on the boundary of this ball.

Now let us continue considering the growth of some other functions in M with respect to $\|F\|^2$.

Let $p \in M$ be a point and $\{V_1, \ldots, V_m\}$ be an orthonormal basis of $T_p M$ and extend it locally so that $\nabla V_i(p) = 0$ for all $i = 1, \ldots, m$, then

$$\sum_{i=1}^m \sqrt{\pm\|(\nabla)^k(A_\pm(V_i,V_i))\|^2} \leq \sum_{ij}\sqrt{\pm\|((\nabla)^k A_\pm)(V_i,V_j)\|^2} = \sqrt{\pm\|(\nabla)^k A_\pm\|^2}.$$

But $\vec{H}$ is the trace of A, so that

$$\sqrt{\pm\|(\nabla)^k \vec{H}_\pm\|^2} = \sqrt{\pm\left\|\sum_{i=1}^m (\nabla)^k [A_\pm(V_i,V_i)]\right\|^2}$$
$$\leq \sum_{i=1}^m \sqrt{\pm\|[(\nabla)^k A_\pm](V_i,V_i)\|^2} \leq \sqrt{\pm\|(\nabla)^k A_\pm\|^2},$$

which implies:

$$-\|(\nabla)^k \vec{H}_-\|^2 \leq -\|(\nabla)^k A_-\|^2 \leq c_k, \tag{4.1}$$

$$\|(\nabla)^k \vec{H}_+\|^2 \leq \|(\nabla)^k A_+\|^2 \leq d_k, \tag{4.2}$$

everywhere in M.

Then the bounded geometry assumption **excludes the mainly negative case** for spacelike self-shrinkers, because $-d_0 \leq \|\vec{H}\|^2 \leq c_0$ and, in this case, $\|F\|^2$ has no lower bound (by Lemma 4.0.11), but $\|F\|^2 = \|F^\top\|^2 + \|F^\perp\|^2$ and $\|F^\top\|^2 \geq 0$, which implies that $\|F^\perp\|^2$ is not bounded below. This contradicts $\|F^\perp\|^2 = \|\vec{H}\|^2$. We then proved:

Theorem 4.0.20. *There are no unbounded mainly negative spacelike self-shrinkers of the mean curvature flow with bounded geometry.*

Although it is necessary that $\|F\|^2 \to +\infty$ note that $\|\vec{H}\|^2$ could, so far, still be negative.

Remark 4.0.21. As a remark to the notation we always take the positive (or negative) projection last. This means, for example, that $F_+^\perp := (F^\perp)_+$.

Until now, we have calculated inequalities for the positive and negative projections of some tensors, but we also need inequalities for whole tensors.

Lemma 4.0.22. *For any $X, Y \in \mathbb{R}^{q,n}$ it holds*

$$|\langle X, Y \rangle| \leq \|X_+\|\|Y_+\| + \sqrt{(-\|X_-\|^2)}\sqrt{(-\|Y_-\|^2)}$$

Proof.

$$|\langle X, Y \rangle| = |\langle X_+, Y_+ \rangle + \langle X_-, Y_- \rangle| \leq |\langle X_+, Y_+ \rangle| + |-\langle X_-, Y_- \rangle|$$
$$\leq \|X_+\|\|Y_+\| + \sqrt{(-\|X_-\|^2)}\sqrt{(-\|Y_-\|^2)}.$$

$\square$

This implies:

Lemma 4.0.23. *If $A, B \in \Gamma(F^*T\mathbb{R}^{q,n} \otimes TM \otimes \ldots \otimes TM \otimes T^*M \otimes \ldots \otimes T^*M)$ are such that $\|A_+\|, \|A_-\|, \|B_+\|, \|B_-\|$ grow polynomially with respect to $\|F\|^2$, then so does $|\langle A, B \rangle|$.*

Proof. Let $p \in M$ be a point and $\{e_1, \ldots, e_m\}$ be an orthonormal basis to T_pM, beyond this let
$$A^{(j_1,\ldots,j_l)}_{(i_1,\ldots,i_k)} e_{j_1} \otimes \ldots \otimes e_{j_l} := A(e_{i_1}, \ldots, e_{i_k})$$
and
$$B^{(j_1,\ldots,j_l)}_{(i_1,\ldots,i_k)} e_{j_1} \otimes \ldots \otimes e_{j_l} := B(e_{i_1}, \ldots, e_{i_k})$$
so that $A^{(j_1,\ldots,j_l)}_{(i_1,\ldots,i_k)}, B^{(j_1,\ldots,j_l)}_{(i_1,\ldots,i_k)} \in F^*T\mathbb{R}^{q,n}(p)$ and
$$|\langle A, B \rangle| = \left| \sum_{i_1,\ldots,i_k,j_1,\ldots,j_l} \left\langle A^{(j_1,\ldots,j_l)}_{(i_1,\ldots,i_k)}, B^{(j_1,\ldots,j_l)}_{(i_1,\ldots,i_k)} \right\rangle \right|$$
$$\leq \sum_{i_1,\ldots,i_k,j_1,\ldots,j_l} \left| \left\langle A^{(j_1,\ldots,j_l)}_{(i_1,\ldots,i_k)}, B^{(j_1,\ldots,j_l)}_{(i_1,\ldots,i_k)} \right\rangle \right|$$
and the conclusion follows immediately with Lemma 4.0.22. □

Remark 4.0.24. At any point $p \in M$,
$$\|F\|^2 = \|\vec{H}\|^2 + \|F^\top\|^2, \tag{4.3}$$
so that, from $|\|\vec{H}\|^2| \leq c_0 + d_0$, it holds that $\|F^\top\|^2$ grows polynomially with respect to $\|F\|^2$.

Lemma 4.0.25. *Let $F : M \to \mathbb{R}^{q,n}$ be a spacelike, mainly positive, immersion with bounded geometry and $f : M \to \mathbb{R}$ be some kind of polynomial (of inner products) of $\vec{H}$, A, their covariant derivatives, F, $F^\top$ and the function $\frac{1}{\|\vec{H}\|}$, then f has polynomial growth with respect to $\|F\|^2$.*

Proof. As M has a positive definite metric, one can use the Cauchy-Schwarz inequality for $F^\top$ and tensors on TM and T^*M and equation (4.3) grants the desired growth for $F^\top$. Beyond this, the Lemma follows from the bounded geometry on $\|(\nabla)^k A\|$, $\|(\nabla)^k \vec{H}\|$ and $\frac{1}{\|\vec{H}\|}$ together with Lemmas 4.0.22 and 4.0.23. □

We will integrate over the whole manifold with respect to the following *heat kernel*: $\rho : M \to \mathbb{R}$ defined as
$$\rho(x) := \exp\left(-\frac{\|F\|^2}{2}\right).$$

Lemma 4.0.26. *Let $F : M \to \mathbb{R}^{q,n}$ be a spacelike, mainly positive, immersion with bounded geometry and $F(M)$ unbounded, beyond this let $f : M \to \mathbb{R}$ be some polynomial (of inner products) of $\vec{H}$, A, their covariant derivatives, F, $F^\top$ and the function $\frac{1}{\|\vec{H}\|}$. Then*

$$\left| \int_M f \rho d\mu \right| < \infty;$$

beyond this, one can use partial integration

$$\int_M \rho \, \mathrm{div}(\nabla f(x)) d\mu = - \int_M \langle \nabla \rho, \nabla f(x) \rangle d\mu.$$

Proof. By Lemma 4.0.25 f has polynomial growth with respect to $\|F\|^2$. Otherwise let $R \in \mathbb{R}$ be big enough so that the mainly positive condition is satisfied (for some $\epsilon > 0$) for all $x \in M$ with $\|F(x)\|_\mathbb{E} > R$ and let $p \in M$ be a fixed point with $\|F(p)\|_\mathbb{E} \leq R$. Then choose $R'_i = 2\sqrt{R+i}/k$ as in Lemma 4.0.15, where k is the inverse Lipschitz constant, so that $\|F(x)\|_\mathbb{E} \geq \sqrt{R+i}$, for all $x \in B_p(R'_i)$ and for all $i = 1, 2, \ldots$ Beyond this, for any $N \in \mathbb{N}$,

$$\left| \int_{B_p(R'_N)} \rho f(x) d\mu \right| \leq \left| \int_{B_p(R'_0)} \rho f(x) d\mu \right| + \sum_{i=1}^N \left| \int_{B_p(R'_i) \setminus B_p(R'_{i-1})} \rho f(x) d\mu \right|.$$

Because of the polynomial growth of f (Lemma 4.0.25) and $\mathrm{vol}(B_p(R))$ (definition 4.0.14) with respect to $\|F(x)\|^2$ and Rem. 4.0.19, it is possible to make R big enough so that, for every $x \in (B_p(R'_i) \setminus B_p(R'_{i-1}))$ and for all $i \in \mathbb{N}$, it holds that

$$\mathrm{vol}(B_p(R'_i) \setminus B_p(R'_{i-1})) \cdot \sup_{B_p(R'_i) \setminus B_p(R'_{i-1})} \rho f(x) \leq \mathcal{P}(\|F(x)\|^2) \exp\left(-\frac{\|F(x)\|^2}{2} \right), \tag{4.4}$$

for some polynomial $\mathcal{P} : \mathbb{R} \to \mathbb{R}$. But, because of the exponential decay of $\exp\left(-\frac{\|F(x)\|^2}{2}\right)$, the right side of this inequality is smaller than any inverse polynomial in $\|F(x)\|^2$, in particular smaller than $\|F(x)\|^{-2}$ whenever $\|F(x)\|^2 \geq Q$, for some $Q \in \mathbb{R}$. On the other hand, because of Cor. 4.0.16 and Rem. 4.0.17, for any $x \in (B_p(R'_i) \setminus B_p(R'_{i-1}))$, it holds that

$$\|F(x)\|^2 \geq \frac{\epsilon}{2} \|F(x)\|_\mathbb{E}^2 \geq \frac{\epsilon}{2}(R+i-1)$$

for every $x \in B_p(R'_i) \setminus B_p(R'_{i-1})$ so that if R is big enough so that $\frac{\varepsilon}{2}R > Q$, the right side of eq. (4.4) is smaller than $\frac{C}{(R+i-1)^2}$, for some $C \in \mathbb{R}$. This implies, for any $N \in \mathbb{N}$,

$$\left|\int_{B_p(R'_N)} \rho f(x) d\mu\right| \leq \left|\int_{B_p(R'_0)} \rho f(x) d\mu\right| + \sum_{i=1}^{N} \frac{C}{(R+i-1)^2}$$
$$\leq \left|\int_{B_p(R'_0)} \rho f(x) d\mu\right| + \sum_{i=1}^{\infty} \frac{C}{(R+i-1)^2} \in \mathbb{R}.$$

Then the bounded convergence Theorem of Lebesgue implies that the sequence $f(x)\chi_{B_p(R'_i)}$, with $\chi_{B_p(R'_i)}$ the characteristic function of the ball $B_p(R'_i)$, converges to an integrable function, which is f.

Using the divergence Theorem with η as the outher normal to $\partial B_p(R)$ it holds that

$$\int_{B_p(R)} \rho \operatorname{div}(\nabla f(x)) d\mu = \int_{\partial B_p(R)} \rho \nabla_\eta f(x) d\mu - \int_{B_p(R)} \langle \nabla \rho, \nabla f(x)\rangle d\mu,$$

but

$$\int_{\partial B_p(R)} \rho \nabla_\eta f(x) d\mu \to 0$$

when $R \to \infty$ through an argument analogous to the first part of this Lemma because $\operatorname{vol}(\partial B_p(R))$ is growing much slower than $\rho \nabla_\eta f(x)$ is going to zero. Thence the claim on partial integration holds. □

By this, all integrals in the next Lemma are finite.

Lemma 4.0.27. *Let $F : M \to \mathbb{R}^{q,n}$ be a spacelike, mainly positive, self-shrinker of the mean curvature flow with bounded geometry such that $F(M)$ is unbounded. Beyond this, let F satisfy $\|\vec{H}\|^2 \neq 0$ and $\nabla^\perp \nu = 0$. Then*

$$\nabla_i \|\vec{H}\| \frac{P_{jk}}{\|\vec{H}\|} - \|\vec{H}\| \nabla_i \left(\frac{P_{jk}}{\|\vec{H}\|}\right) = 0.$$

Proof. From equation (3.2.5) we have

$$\int_M \rho \frac{\|P\|^2}{\|\vec{H}\|^2} \Delta \left(\frac{\|P\|^2}{\|\vec{H}\|^4}\right) d\mu = \int_M \rho \frac{\|P\|^2}{\|\vec{H}\|^2} \frac{2}{\|\vec{H}\|^4} \left\| \nabla_i \|\vec{H}\| \frac{P_{jk}}{\|\vec{H}\|} - \|\vec{H}\| \nabla_i \left(\frac{P_{jk}}{\|\vec{H}\|}\right) \right\|^2$$
$$+ \rho \frac{\|P\|^2}{\|\vec{H}\|^2} \left\langle F^T, \nabla \left(\frac{\|P\|^2}{\|\vec{H}\|^4}\right) \right\rangle$$
$$- \rho \frac{\|P\|^2}{\|\vec{H}\|^2} \frac{2}{\|\vec{H}\|} \left\langle \nabla \|\vec{H}\|, \nabla \left(\frac{\|P\|^2}{\|\vec{H}\|^4}\right) \right\rangle d\mu,$$

but, using partial integration[3], we have the following

$$\int_M \rho \frac{\|P\|^2}{\|\vec{H}\|^2} \Delta \left(\frac{\|P\|^2}{\|\vec{H}\|^4}\right) d\mu = -\int_M \left\langle \nabla \rho \frac{\|P\|^2}{\|\vec{H}\|^2}, \nabla \left(\frac{\|P\|^2}{\|\vec{H}\|^4}\right) \right\rangle$$
$$+ \rho \left\langle \nabla \left(\frac{\|P\|^2}{\|\vec{H}\|^4}\right) \|\vec{H}\|^2, \nabla \left(\frac{\|P\|^2}{\|\vec{H}\|^4}\right) \right\rangle + \rho \frac{\|P\|^2}{\|\vec{H}\|^4} \left\langle \nabla \|\vec{H}\|^2, \nabla \left(\frac{\|P\|^2}{\|\vec{H}\|^4}\right) \right\rangle d\mu$$
$$= -\int_M -\rho \frac{\|P\|^2}{\|\vec{H}\|^2} \left\langle F^T, \nabla \left(\frac{\|P\|^2}{\|\vec{H}\|^4}\right) \right\rangle + \rho \|\vec{H}\|^2 \left\| \nabla \left(\frac{\|P\|^2}{\|\vec{H}\|^4}\right) \right\|^2$$
$$+ \rho \frac{\|P\|^2}{\|\vec{H}\|^2} \frac{2}{\|\vec{H}\|} \left\langle \nabla \|\vec{H}\|, \nabla \left(\frac{\|P\|^2}{\|\vec{H}\|^4}\right) \right\rangle d\mu$$

so that, equating the two equations for $\int_M \rho \frac{\|P\|^2}{\|\vec{H}\|^2} \Delta \left(\frac{\|P\|^2}{\|\vec{H}\|^4}\right) d\mu$,

$$\int_M 2\rho \frac{\|P\|^2}{\|\vec{H}\|^6} \left\| \nabla_i \|\vec{H}\| \frac{P_{jk}}{\|\vec{H}\|} - \|\vec{H}\| \nabla_i \left(\frac{P_{jk}}{\|\vec{H}\|}\right) \right\|^2 + \rho \|\vec{H}\|^2 \left\| \nabla \left(\frac{\|P\|^2}{\|\vec{H}\|^4}\right) \right\|^2 d\mu = 0. \tag{4.5}$$

but $\|\vec{H}\|^2 \neq 0$, so that $\|\vec{H}\|^2$ has the same sign everywhere and so have the summands inside the integral. This implies in particular, using $\|P\|^2 \neq 0$ (because $P = 0$ would imply $\|\vec{H}\| = 0$), that

$$\left\| \nabla_i \|\vec{H}\| \frac{P_{jk}}{\|\vec{H}\|} - \|\vec{H}\| \nabla_i \left(\frac{P_{jk}}{\|\vec{H}\|}\right) \right\|^2 = 0$$

---

[3]One could use $\nabla^i \left(\frac{\|P\|^2}{\|\vec{H}\|^2} \nabla_i \left(\frac{\|P\|^2}{\|\vec{H}\|^4}\right)\right) = \nabla^i \left(\frac{\|P\|^2}{\|\vec{H}\|^2}\right) \nabla_i \left(\frac{\|P\|^2}{\|\vec{H}\|^4}\right) + \frac{\|P\|^2}{\|\vec{H}\|^2} \Delta \left(\frac{\|P\|^2}{\|\vec{H}\|^4}\right)$ to put the first expression in the form of Lemma 4.0.26.

and, as M is spacelike,
$$\nabla_i \|\vec{H}\| \frac{P_{jk}}{\|\vec{H}\|} - \|\vec{H}\| \nabla_i \left(\frac{P_{jk}}{\|\vec{H}\|}\right) = 0.$$

$\square$

And we follow exactly as in the compact case.

Lemma 4.0.28. *Let M be a smooth manifold and $F : M \to \mathbb{R}^{q,n}$ be mainly positive, spacelike, self-shrinker of the mean curvature flow with bounded geometry such that $F(M)$ is unbounded, beyond that let F satisfy the conditions: $\|\vec{H}\|^2(p) \neq 0$ for all $p \in M$ and the principal normal is parallel in the normal bundle ($\nabla^\perp \nu \equiv 0$). Then one of the two holds*

1. $\nabla \|\vec{H}\| = 0$ *everywhere on M*

2. $\frac{\nabla \|\vec{H}\|}{\|\nabla \|\vec{H}\|\|}$ *is the only eigenvector associated with a nonzero eigenvalue of P.*

Proof. First, from Lemma 4.0.27
$$* := \nabla_i \|\vec{H}\| \frac{P_{jk}}{\|\vec{H}\|} - \|\vec{H}\| \nabla_i \left(\frac{P_{jk}}{\|\vec{H}\|}\right) = 0. \tag{4.6}$$

Second, using the Codazzi equation (eq. (1.3.1)) and $\nabla^\perp \nu = 0$, we calculate
$$\nabla_i \left(\frac{P_{jk}}{\|\vec{H}\|}\right) = \nabla_i \langle \nu, A_{jk} \rangle = \langle \nu, \nabla_i^\perp A_{jk} \rangle$$
$$= \langle \nu, \nabla_j^\perp A_{ik} \rangle = \nabla_j \langle \nu, A_{ik} \rangle = \nabla_j \left(\frac{P_{ik}}{\|\vec{H}\|}\right).$$

Third, using equation (3.25), we write
$$* = \left(\nabla_i \|\vec{H}\| \frac{P_{jk}}{\|\vec{H}\|} - \nabla_j \|\vec{H}\| \frac{P_{ik}}{\|\vec{H}\|}\right) + \left(\nabla_j \|\vec{H}\| \frac{P_{ik}}{\|\vec{H}\|} - \|\vec{H}\| \nabla_i \left(\frac{P_{jk}}{\|\vec{H}\|}\right)\right)$$
$$= \left(\nabla_i \|\vec{H}\| \frac{P_{jk}}{\|\vec{H}\|} - \nabla_j \|\vec{H}\| \frac{P_{ik}}{\|\vec{H}\|}\right) + \left(\nabla_j \|\vec{H}\| \frac{P_{ik}}{\|\vec{H}\|} - \|\vec{H}\| \nabla_j \left(\frac{P_{ik}}{\|\vec{H}\|}\right)\right)$$
$$= \left(\nabla_i \|\vec{H}\| \frac{P_{jk}}{\|\vec{H}\|} - \nabla_j \|\vec{H}\| \frac{P_{ik}}{\|\vec{H}\|}\right),$$

which implies

$$0 = \left\| \nabla_i \|\vec{H}\| \frac{P_{jk}}{\|\vec{H}\|} - \|\vec{H}\| \nabla_i \left(\frac{P_{jk}}{\|\vec{H}\|} \right) \right\|^2 = \left\| \nabla_i \|\vec{H}\| \frac{P_{jk}}{\|\vec{H}\|} - \nabla_j \|\vec{H}\| \frac{P_{ik}}{\|\vec{H}\|} \right\|^2.$$

Now, expanding this norm we find that

$$\|\nabla \|\vec{H}\| \|^2 \frac{\|P\|^2}{\|\vec{H}\|^2} - \frac{\nabla_i \|\vec{H}\|}{\|\vec{H}\|^2} P_{jk} \nabla^j \|\vec{H}\| P^{ik} = 0,$$

$$\|\nabla \|\vec{H}\| \|^2 \|P\|^2 - \|P(\nabla \|\vec{H}\|)\|^2 = 0. \tag{4.7}$$

We now calculate at a point $p \in M$ fixed. As the 2-tensor P is symmetric, it is also diagonalizable and has only real eigenvalues $\lambda_1, \ldots, \lambda_m$. Let $V_1, \ldots, V_m$ be an orthonormal basis of eigenvectors associated with $\lambda_1, \ldots, \lambda_m$. Then we write $\nabla \|\vec{H}\| = \Sigma_i \alpha_i V_i$, $\alpha_i \in \mathbb{C}$, so that by equation (4.7)

$$0 = \|P\|^2 \|\nabla \|\vec{H}\| \|^2 - \|P(\nabla \|\vec{H}\|)\|^2 = \sum_i \lambda_i^2 (\|\nabla \|\vec{H}\| \|^2 - \alpha_i^2), \tag{4.8}$$

but $\lambda_i^2 \geq 0$ because $\lambda_i \in \mathbb{R}$, beyond this $\|\vec{H}\|$ is pure real or pure imaginary everywhere and all the α_i's have to agree with $\|\vec{H}\|$ about being real or imaginary, which implies that

$$\|\nabla \|\vec{H}\| \|^2 - \alpha_i^2 = \sum_{j \neq i} \alpha_j^2$$

is nonnegative for all $i \in \{1, \ldots, m\}$ if $\|\vec{H}\|$ is real and nonpositive for all $i \in \{1, \ldots, m\}$ if $\|\vec{H}\|$ is imaginary. This implies, with eq. (4.8), that

$$\lambda_i^2 (\|\nabla \|\vec{H}\| \|^2 - \alpha_i^2) = 0 \ \forall i \in \{1, \ldots, m\}.$$

As $tr(P) = P_{ij} g^{ij} = \|\vec{H}\|^2 \neq 0$, it follows that $P \neq 0$ and there is at least a $j \in \{1, \ldots, n\}$ such that $\lambda_j \neq 0$ and the last equation implies that

$$0 = \|\nabla \|\vec{H}\| \|^2 - \alpha_j^2 = \sum_i \alpha_i^2 - \alpha_j^2 = \sum_{i \neq j} \alpha_i^2 \implies \alpha_i = 0 \ \forall i \neq j.$$

From this, it follows that $\|\nabla \|\vec{H}\| \|^2 = \alpha_j^2$ and $\nabla \|\vec{H}\| = \alpha_j V_j$.

If there is an $x \in M$ such that $\nabla \|\vec{H}\|(x) \neq 0$, then $\alpha_j \neq 0$ and for all $i \neq j$

$$0 = \lambda_i^2(\|\nabla\|\vec{H}\|\|^2 - \alpha_i^2) = \lambda_i^2 \alpha_j^2 \implies \lambda_i = 0,$$

so that P_{ij} has only one nonzero eigenvalue and the associated eigenvector is $\nabla\|\vec{H}\|/\|\nabla\|\vec{H}\|\|$. □

The rest of the proof in the compact case cannot be extended to the non-compact case, so that we are forced to consider two different possibilities:

1. $\nabla\|\vec{H}\| = 0$ everywhere on M
2. There is a point $p \in M$ with $\nabla\|\vec{H}\|(p) \neq 0$, at which $\frac{\nabla\|\vec{H}\|}{\|\nabla\|\vec{H}\|\|}$ is the only eigenvector associated with a nonzero eigenvalue of P.

We have to treat these two cases separately.

4.1 The First Case

The proof of the following Theorem is relatively extensive, it is internally divided into several Lemmas for the comfort of the reader.

As a remark to the notation: Throughout the proof of the next Theorem, we identify vectors, vectorfields and some curves in different manifolds through the immersions among then several times without explicit mention in order to avoid the notation becoming too heavy.

Theorem 4.1.1. *Let M be a smooth manifold and $F : M \to \mathbb{R}^{q,n}$ be a mainly positive, spacelike, shrinking self-similar solution of the mean curvature flow with bounded geometry such that $F(M)$ is unbounded. Beyond that, let F satisfy the conditions: $\|\vec{H}\|^2(p) \neq 0$, $\forall p \in M$, and the principal normal is parallel in the normal bundle ($\nabla^\perp \nu \equiv 0$). If $\nabla\|\vec{H}\|(p) = 0$ for all $p \in M$, then*

$$F(M) = \mathcal{H}_r \times \mathbb{R}^{m-r}, \tag{4.9}$$

where $\mathcal{H}_r$ is an r-dimensional minimal surface of the hyperquadric $\mathcal{H}^{n-1}(r)$ with $\|\vec{H}\|^2 = r > 0$ and $\mathbb{R}^{m-r}$ is an $m-r$ dimensional spacelike affine space in $\mathbb{R}^{q,n}$.

Proof. First we see that $\nabla \|\vec{H}\| = 0$ implies $\nabla^\perp \vec{H} = \nabla\|\vec{H}\|\nu = 0$ and it follows, from equation (3.7), that
$$\theta^i A_{ij} = 0. \tag{4.10}$$
On the other hand, $\nabla \|\vec{H}\| = 0$ implies that $\|\vec{H}\|^2$ is constant, so that, with Lemma 4.0.27, it holds $\nabla P = 0$ and then equation (3.8) implies
$$\langle \nabla_i^\perp \nabla_j^\perp \vec{H}, \vec{H}\rangle = \langle A_{ij} - P_i^k A_{kj} + \theta^k \nabla_i^\perp A_{jk}, \vec{H}\rangle$$
$$0 = P_{ij} - P_i^k P_{kj}, \tag{4.11}$$
so that $P = P^2$; this means P is a projection and the only possible eigenvalues of P are 1 and 0.

First, consider the multiplicity of the eigenvalues of P. Let us write $r(p)$ for the multiplicity of the eigenvalue 1 at a point $p \in M$. Because of $\nabla_k P_{ij} = 0$ we get
$$\nabla_k \|P\|^2(p) = 2\nabla_k P_{ij} P^{ij} = 0,$$
but $\|P\|^2(p) = r(p)$ (because the eigenvalues of P are only 0 or 1.) which implies that r is constant. So that
$$\|\vec{H}\|^2 = r > 0. \tag{4.12}$$

We consider the eigenspaces associated with these two eigenvalues, they define the distributions $\mathcal{E}M$ and $\mathcal{F}M$ given at and point $p \in M$ by
$$\mathcal{E}M_p := \{V \in T_pM : P(V) = V\}, \quad \mathcal{F}M_p := \{V \in T_pM : P(V) = 0\}, \tag{4.13}$$
or in local coordinates $P_i^j V^i = V^j$ (and $P_i^j V^i = 0$). As the eigenspaces are orthogonal we have $T_pM = \mathcal{E}M_p \oplus \mathcal{F}M_p$. From equation (4.10) we have, for all $V \in \mathcal{E}_pM$, that
$$\theta(V) = \theta_j V^j = \theta_j P_i^j V^i = 0, \tag{4.14}$$
which means that $\mathcal{E}M_p \subset \ker(\theta)$.

Lemma 4.1.2. *Under the hypothesis of Theorem 4.1.1 the distributions $\mathcal{E}M$ and $\mathcal{F}M$ are involutive[4].*

---
[4] For involutive see definition 6.2.3.

Proof. For $e_1, e_2 \in \Gamma(\mathcal{E}M)$ and $f_1, f_2 \in \Gamma(\mathcal{F}M)$, from $\nabla P = 0$, we get

$$P(\nabla_{e_1} e_2) = \nabla_{e_1} P(e_2) = \nabla_{e_1} e_2, \tag{4.15}$$
$$P(\nabla_{f_1} f_2) = \nabla_{f_1} P(f_2) = \nabla_{f_1} 0 = 0. \tag{4.16}$$

i.e. $\nabla_{e_1} e_2 \in \Gamma(\mathcal{E}M)$ and $\nabla_{f_1} f_2 \in \Gamma(\mathcal{F}M)$. As the Levi-Civita connection is torsion free we have

$$[e_1, e_2] = \nabla_{e_1} e_2 - \nabla_{e_2} e_1 \in \Gamma(\mathcal{E}M), \tag{4.17}$$
$$[f_1, f_2] = \nabla_{f_1} f_2 - \nabla_{f_2} f_1 \in \Gamma(\mathcal{F}M). \tag{4.18}$$

So $\mathcal{E}M$ and $\mathcal{F}M$ are involutive. $\square$

By the Theorem of Frobenius (Theorem 6.2.4), these distributions define two foliations, such that, at each $p \in M$, there are two leaves $\mathcal{E}_p$ and $\mathcal{F}_p$ that intersect orthogonally at p. We want to understand what they are. The inclusions of these leaves in M, which we denote by $i_{\mathcal{E}_p}$ and $i_{\mathcal{F}_p}$, are immersions (one sees this immediately remembering that the charts on the leaves are induced by the charts on M) and one can draw the following diagrams:

In order to understand what the image of the leaves $\mathcal{E}_p$ and $\mathcal{F}_p$ in $\mathbb{R}^{q,n}$ are, we need further information on the second fundamental tensor of F. To understand it better we need an auxiliary tensor:

Let us define the tensor

$$(P * A)_{ij} := P_i^k A_{kj},$$

although this definition is local, it is globally well defined, because contractions of tensors are tensors. This is a symmetric tensor (by Lemma 3.2.3).

Lemma 4.1.3. *Under the hypothesis of Theorem 4.1.1, the following equations* hold for the second fundamental tensor of F:

$$\theta^k \nabla_k^\perp A_{ij} = 0, \tag{4.19}$$

$$A_{ij} = P_i^k A_{kj}. \tag{4.20}$$

Proof. First, from (3.8) (with $\nabla \vec{H} = 0$) and (4.11), we get

$$\theta^k \nabla_k^\perp (P_i^l A_{lj}) = P_i^l \theta^k \nabla_k^\perp A_{lj} = P_i^l P_l^m A_{mj} - P_i^l A_{lj} =$$
$$\theta^k \nabla_k^\perp (P_i^l A_{lj}) = P_i^l A_{lj} - P_i^l A_{lj} = 0. \tag{4.21}$$

We claim that to prove (4.20), it is enough to show

$$\|A_\pm\|^2 = \|P * A_\pm\|^2.$$

One sees this using eq. (4.11) and calculating

$$\|A_\pm - P * A_\pm\|^2 = \|A_\pm\|^2 + \|P * A_\pm\|^2 - 2\langle P_{il}(A_\pm)_j^l, A_\pm^{ij}\rangle$$
$$= \|A_\pm\|^2 + \|P * A_\pm\|^2 - 2\langle P_{tl}(A_\pm)_j^l, P_i^t A_\pm^{ij}\rangle$$
$$\|A_\pm - P * A_\pm\|^2 = \|A_\pm\|^2 - \|P * A_\pm\|^2,$$

so that $A = P * A \Leftrightarrow \|A_\pm - P * A_\pm\|^2 = 0 \Leftrightarrow \|A_\pm\|^2 = \|P * A_\pm\|^2$.

Let us then prove that $\|A_\pm\|^2 = \|P * A_\pm\|^2$. First of all, we see, using (3.8) (with $\nabla_i \vec{H} = 0$), (4.11) and $\theta^k \nabla_k^\perp (P_i^l A_{lj}) = 0$, that

$$\theta^k \nabla_k (\|A_\pm\|^2 - \|P * A_\pm\|^2) = 2\langle A_\pm^{ij}, \theta^k \nabla_k^\perp (A_\pm)_{ij}\rangle = 2\langle A_\pm^{ij}, P_i^k (A_\pm)_{kj} - (A_\pm)_{ij}\rangle$$
$$= 2\langle A_\pm^{ij}, P_i^l P_l^k (A_\pm)_{kj} - (A_\pm)_{ij}\rangle$$
$$\theta^k \nabla_k (\|A_\pm\|^2 - \|P * A_\pm\|^2) = -2(\|A_\pm\|^2 - \|P * A_\pm\|^2). \tag{4.22}$$

If $\theta = 0$ at some point $p \in M$, then this equation implies $\|A_\pm\|^2 = \|P * A_\pm\|^2$ and $A_{ij} = P_i^l A_{lj}$ at this point. So, without loss of generality, we can consider only the points $q \in M$ with $\theta(q) \neq 0$. Fix one of these and consider the integral curve $\gamma : (-a, b) \to M$ of θ (the vector field identified with this 1-form through the metric[5]) with $\gamma(0) = q$, for some $a, b > 0$. Along this curve we define the

---
[5]It is often written $\theta^\sharp$ in the literature, but we write just θ here, because it is expressed, in local coordinates, as θ^i while the 1-form θ is expressed θ_i.

function
$$f(s) := \|\theta\|^2(\gamma(s)),$$
and get
$$\frac{d}{ds}f = \nabla_{\dot\gamma}\|\theta\|^2 = \theta^k \nabla_k \|\theta\|^2 = 2\theta^k \theta^l \nabla_k \theta_l.$$
but, from $\vec{H} = -F^\perp$,
$$\nabla_i \theta_j = \nabla_i \langle F, F_j \rangle = g_{ij} - \langle \vec{H}, A_{ij} \rangle$$
and $\theta^i \nabla_i \theta_j = \theta_j$ because of equation (4.10), so that
$$\frac{d}{ds}f = 2f. \tag{4.23}$$
This has a unique solution with $f(0) = \|\theta\|^2(q)$:
$$\|\theta\|^2(\gamma(s)) = \|\theta\|^2(q) e^{2s} > 0,$$
in particular $\|\theta\|^2(\gamma(s)) \neq 0$ for all $s \in (a,b)$, then these integral curves do not cross any singular point and the maximal integral curve is defined for all $\mathbb{R}$ ($a = -\infty$, $b = \infty$) and it is not closed (because of injectivity of e^{2s}). Over this same curve we define functions $\tilde{f}_\pm : \mathbb{R} \to \mathbb{R}$,
$$\tilde{f}_\pm(s) := (\|A_\pm\|^2 - \|P * A_\pm\|^2)(\gamma(s))$$
and, using equation (4.22), get
$$\begin{aligned}\frac{d\tilde{f}_\pm}{ds} &= \dot\gamma(\|A_\pm\|^2 - \|P * A_\pm\|^2) = \theta^k \nabla_k (\|A_\pm\|^2 - \|P * A_\pm\|^2) \\ &= -2(\|A_\pm\|^2 - \|P * A_\pm\|^2) = -2\tilde{f}_\pm.\end{aligned}$$
This has a unique solution with $\tilde{f}_\pm(0) = (\|A_\pm\|^2 - \|P * A_\pm\|^2)(q)$:
$$(\|A_\pm\|^2 - \|P * A_\pm\|^2)(\gamma(s)) = (\|A_\pm\|^2 - \|P * A_\pm\|^2)(q) e^{-2t}.$$
If $(\|A_\pm\|^2 - \|P * A_\pm\|^2)(q) \neq 0$, then $(\|A_\pm\|^2 - \|P * A_\pm\|^2)(\gamma(s)) \to \pm\infty$ as $s \to -\infty$ and this contradicts the boundedness of the second fundamental tensor

(from the bounded geometry condition). So, $\|A_\pm\|^2 - \|P * A_\pm\|^2 = 0$ and $A = P * A$. Then equation (4.21) implies (4.19). $\square$

Let us now examine the leaves of the distribution $\mathcal{E}M$.

Lemma 4.1.4. *Under the hypothesis of Theorem 4.1.1 it holds that $\mathcal{E}_p$ is immersed into $\mathcal{H}^{n-1}(\|F\|^2(p))$ through $F \circ i_{\mathcal{E}_p}$. Beyond this, $\mathcal{E}_p$ is geodesically complete and there is $q \in M$ so that $F \circ i_{\mathcal{E}_q}$ is a minimal immersion into $\mathcal{H}^{n-1}(\|F\|^2(q))$.*

Proof. First of all, $\mathcal{E}_p$ is an r-dimensional manifold immersed in M under the natural inclusion $i_{\mathcal{E}_p} : \mathcal{E}_p \to M$, then $F \circ i_{\mathcal{E}_p} : \mathcal{E}_p \to \mathbb{R}^{q,n}$ is also immersion and we can consider the second fundamental tensors of them:

$$\begin{array}{ccc} M & \xrightarrow{F} & \mathbb{R}^{q,n} \\ i_{\mathcal{E}_p} \uparrow & \nearrow_{F \circ i_{\mathcal{E}_p}} & \\ \mathcal{E}_p & & \end{array}$$

Let us drop the index p to make the notation not so heavy. So, $A_{F \circ i_\mathcal{E}}$ and $A_{i_\mathcal{E}}$ denote the second fundamental tensors of $F \circ i_\mathcal{E}$ and $i_\mathcal{E}$ respectively. From equation (2.4), identifying vector fields over $\mathcal{E}_p$ and in M through $di_\mathcal{E}$, it holds that

$$A_{F \circ i_\mathcal{E}} = A_F + dF(A_{i_\mathcal{E}}). \tag{4.24}$$

On the other hand one can write[6], for local vector fields $e_1, e_2 \in \Gamma(T\mathcal{E}_p)$, the second fundamental tensor of the immersion $i_\mathcal{E}$ as

$$A_{i_\mathcal{E}}(e_1, e_2) = \nabla_{e_1} e_2 - \nabla'_{e_1} e_2, \tag{4.25}$$

where ∇' is the Levi-Civita connection of $\mathcal{E}_p$ (with respect to metric induced by the inclusion) and we are identifying vectors in $T_q\mathcal{E}$ with vectors in T_qM through $di_{\mathcal{E}_p}$[7]. But $\nabla_{e_1} e_2 \in \Gamma(\mathcal{E}M)$ by eq. (4.15) and $di_\mathcal{E}(\nabla'_{e_1} e_2) \in \Gamma(\mathcal{E}M)$ by definition, but $A_{i_\mathcal{E}}(e_1, e_2) \in \Gamma(T\mathcal{E}^\perp)$, so that equation (4.25) implies

$$A_{i_\mathcal{E}} = 0, \tag{4.26}$$

---

[6] One can easily see that this agrees with the previous definition of the second fundamental tensor in local coordinates.
[7] For more details on this a reference is [dC92].

which means (by definition) that $i_\mathcal{E}$ is a totally geodesic immersion. This means, in particular, that the (image through $i_\mathcal{E}$ of the) geodesics of $\mathcal{E}_p$ are geodesics of M too, because eq. (4.26) with eq. (4.25) implies that $\nabla = \nabla'$.

Let us now show that $\mathcal{E}_p$ is a complete manifold. This is done in the following Note to easily cite this same argument in other parts of the work.

Note 4.1.5. Let $\gamma : (-a,b) \to \mathcal{E}_p$, $a,b > 0$, be a maximally extended geodesic of $\mathcal{E}_p$ (the interval is open because of the definition of a leaf, as a union of plaques), we set $\delta := i_\mathcal{E} \circ \gamma$. Then $0 = \nabla'_{\dot\gamma} \dot\gamma = \nabla_{\dot\delta} \dot\delta$ implies that δ is a geodesic of M. As M is geodesically complete, δ can be infinitely extended, $\delta : \mathbb{R} \to M$. By contradiction let us assume that $b < \infty$ (the case $a \neq \infty$ is analogous). As the tensor P is continuous and $P(\dot\delta(t)) = \dot\delta(t)$ for all $t \in (-a,b)$, it holds that $P(\dot\delta(b)) = \dot\delta(b)$ i. e. $\dot\delta(b) \in \mathcal{E}M$. Now consider the leaf $\mathcal{E}_{\delta(b)}$. Let $i_{\mathcal{E}(\delta(b))}$ be the immersion of this leaf in M and $p_0 \in (i_{\mathcal{E}(\delta(b))})^{-1}(\delta_b)$, then its differential $di_{\mathcal{E}(\delta(b))} : T_{p_0}\mathcal{E}_{\delta(b)} \to \mathcal{E}M_{\delta(b)}$ is bijective. Let $V := (di_{\mathcal{E}(\delta(b))})^{-1} \dot\delta(b)$ and $\gamma_0 : (-c,d) \to \mathcal{E}_{\delta(b)}$ be the geodesic with $\gamma_0(0) = p_0$ and $\dot\gamma_0(0) = V$. Then $\delta_0 := i_{\mathcal{E}(\delta(b))} \circ \gamma_0$ is the geodesic of M with $\delta_0(0) = \delta(b)$ and $\dot\delta_0(0) = \dot\delta(b)$, so that $\delta_0(t) = \delta(t+b)$. This means that $\delta(b-\epsilon) \in i_{\mathcal{E}(\delta(b))}(\mathcal{E}_{\delta(b)})$ for all $\epsilon \in (0,c)$. This way $\mathcal{E}_{\delta(b)}$ and $\mathcal{E}_p$ are the same leaf and this contradicts the maximality of $b < \infty$. So, $a = b = \infty$ and $\mathcal{E}_p$ is geodesically complete.

From equation (4.14) we get, for any $q \in \mathcal{E}_p$ and all $V \in \mathcal{E}M_{i_\mathcal{E}(q)}$, $V = V^i \frac{\partial}{\partial x^i}$,

$$0 = \theta_j V^j = \langle F, F_j \rangle V^j = \langle F, dF(V) \rangle,$$

which means that $F(i_\mathcal{E}(q)) \in T_q\mathcal{E}_p^\perp$ (the normal bundle of the immersion $F \circ i_\mathcal{E}$) and

$$V\|F\|^2 = 2V^j \langle F_j, F \rangle = 2\langle dF(V), F \rangle = 0, \qquad (4.27)$$

so that $\|F\|^2$ is constant on the leaf $\mathcal{E}_p$ (but it depends on p), and $F \circ i_\mathcal{E}(\mathcal{E}_p)$ is contained in the hyperquadric $\mathcal{H}^{n-1}(\|F\|^2(p))$. As the inclusion of $\mathcal{H}^{n-1}(\|F\|^2(p))$ in $(\mathbb{R}^n, \langle \cdot, \cdot \rangle)$ is a bijective immersion on its image, it holds that $\mathcal{E}_p$ is immersed in the hyperquadric $\mathcal{H}^{n-1}(\|F\|^2(p))$.

Let us now take a look at a special leaf of the distribution $\mathcal{E}M$. Because of Remark 4.0.17, we have that $\|F\|^2 \geq \mathcal{E}/2\|F\|_\mathbb{E}^2 > k_1$ outside the euclidean sphere $S^{n-1}(2k_1/\mathcal{E})$ but, by Lemma 4.0.15, the inverse image of $\{X \in \mathbb{R}^{q,n} : \|X\|_\mathbb{E}^2 < 2k_1/\mathcal{E}\}$ through F is contained in a geodesic ball of M. Let $k_1 > \inf_{x \in M} \|F(x)\|^2$, then $\inf_{x \in M} \|F(x)\|^2$ must be assumed by some point inside the geodesic ball, i. e. there is a point $q \in M$, such that $\|F(q)\|^2 = \min_{x \in M} \|F(x)\|^2$. Let us consider the leaf $\mathcal{E}_q$. We choose this "smallest" leaf to prove that it is a minimal surface of the hyperquadric. At the end of this Theorem we get that $F(M)$ is some kind of cylinder. The figure 4.2 shows the intersection of a cylinder with two spheres, the small circle in the middle is a minimal surface of the smallest sphere but the two other circles are not a minimal surfaces of the bigger sphere.

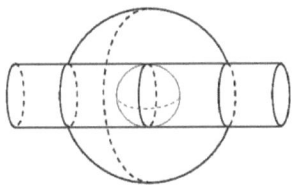

Figure 4.2: Intersection of a cylinder with spheres

The norm of F must be constant over this leaf by eq. (4.27), so that all the points of the leaf minimize the norm of F. But then, $2\langle dF(X), F \rangle = X\|F\|^2 = 0$ for any $X \in T_{q'}M$, $q' \in i_{\mathcal{E}}(\mathcal{E}_q)$, this means that $F(q')$ is orthogonal to $T_{q'}M$, i. e.

$$F^\perp(q') = F(q') \tag{4.28}$$

for every $q' \in \mathcal{E}_q$.

We claim that $F \circ i_{\mathcal{E}}(\mathcal{E}_q)$ is a minimal surface of the hyperquadric $\mathcal{H}^{n-1}(\|F\|^2(q))$. First, the Levi-Civita connection on the hyperquadric is given by the projection ($\Pr_{\mathcal{H}^{n-1}}$) of the Levi-Civita connection of $\mathbb{R}^{q,n}$, which we denote D, into the tangent bundle of the hyperquadric. The following holds for the second fundamental tensor $A_{\mathcal{H}^{n-1}}$ of the immersion of $\mathcal{E}_q$ in $\mathcal{H}^{n-1}(\|F\|^2(q))$ for any $X, Y \in T_x \mathcal{E}_q$

$$\begin{aligned} A_{\mathcal{H}^{n-1}}(X,Y) &= \Pr_{\mathcal{H}^{n-1}}(D_X Y) - \nabla'_X Y = \Pr_{\mathcal{H}^{n-1}}(D_X Y - \nabla'_X Y) \\ A_{\mathcal{H}^{n-1}}(X,Y) &= \Pr_{\mathcal{H}^{n-1}}(A_{F \circ i_\mathcal{E}}(X,Y)) = \Pr_{\mathcal{H}^{n-1}}(A_F(X,Y)), \end{aligned} \tag{4.29}$$

where, in the last step, we used eq. (4.24) with $A_{i_\mathcal{E}} = 0$ (eq. (4.26)).

On the other hand, let us take a vector $V \in T_{q'}M^\perp$, $q' \in \mathcal{E}_q$, that is orthogonal to $\vec{H}$, then, using that $P^{ij}A_{ij}$ is in the same direction as $\vec{H}$ (from Lemma 3.2.3),

one gets
$$\langle P^{ij} A_{ij}, V \rangle = 0.$$

Let us now take an orthonormal basis $\{e_1, \ldots, e_r, f_1, \ldots, f_{m-r}\}$ of $T_{q'}M$ such that $\{e_1, \ldots, e_r\}$ is a basis of $\mathcal{E}M_{q'}$ and $\{f_1, \ldots, f_{m-r}\}$ is a basis of $\mathcal{F}M_{q'}$ then

$$\operatorname{tr}_{\mathcal{E}}\langle A, V \rangle = \sum_{i=1}^{r} \langle V, A(e_i, e_i) \rangle$$
$$= \sum_{i=1}^{r} \langle V, A(P(e_i), e_i) \rangle + \sum_{i=1}^{m-r} \langle V, A(P(f_i), f_i) \rangle = \operatorname{tr}_M \langle V, P*A \rangle = 0,$$

where we used that $P(e_i) = e_i$ and $P(f_i) = 0$. This holds for any $q' \in i_{\mathcal{E}}(\mathcal{E}_q)$ and means that $\operatorname{tr}_{\mathcal{E}} A = \mathfrak{a}(x)\vec{H}$ for some continuous function $\mathfrak{a}: \mathcal{E}_q \in \mathbb{R}$. With this, using eq. (4.28), eq. (4.29) and denoting $\vec{H}_{\mathcal{H}^{n-1}}$ the mean curvature vector of the immersion of $\mathcal{E}_q$ into $\mathcal{H}^{n-1}(\|F\|^2(q))$, we get at q'

$$\vec{H}_{\mathcal{H}^{n-1}} = \Pr_{\mathcal{H}^{n-1}}(\operatorname{tr}_{\mathcal{E}} A) = \Pr_{\mathcal{H}^{n-1}}(\mathfrak{a}\vec{H}) = \Pr_{\mathcal{H}^{n-1}}(-\mathfrak{a} F^{\perp}) = \Pr_{\mathcal{H}^{n-1}}(-\mathfrak{a}(x)F) = 0,$$

because the position vector is orthogonal to the hyperquadric. This means that $\mathcal{E}_q$ is a minimal surface of the hyperquadric $\mathcal{H}^{n-1}(r)$ because $\|F\|^2(q) = \|\vec{H}\|^2(q) = r$ by eq. (4.12). $\square$

We will now analyse the leaves of the distribution $\mathcal{F}M$.

Lemma 4.1.6. *Under the hypothesis of Theorem 4.1.1, it holds that $F \circ i_{\mathcal{F}}(\mathcal{F}_p)$ is an affine space in $\mathbb{R}^{q,n}$ for any $p \in M$. Beyond that, if $q \in i_{\mathcal{E}}(\mathcal{E}_p)$, then $F \circ i_{\mathcal{F}}(\mathcal{F}_p)$ and $F \circ i_{\mathcal{F}}(\mathcal{F}_q)$ are parallel.*

Proof. First, we show that they are affine subspaces of $\mathbb{R}^{q,n}$. Let $q \in i_{\mathcal{F}}(\mathcal{F}_p)$ be an arbitrary point and $f \in \mathcal{F}M_q$ and $X \in T_q M$ be vectors, then equation (4.20) ($A_{ij} = P_i^k A_{kj}$) implies

$$A(f, X) = X^j f^i A_{ij} = X^j f^i P_i^k A_{kj} = 0 \qquad (4.30)$$

because $\mathcal{F}M_q$ is formed by the vectors in the null space of P.

It is clear that the natural inclusion $i_{\mathcal{F}_p}: \mathcal{F}_p \to M$ is an immersion, so that $F \circ i_{\mathcal{F}_p}: \mathcal{F}_p \to \mathbb{R}^{q,n}$ is also an immersion and we can consider the second fundamental tensors of them:

$$\begin{array}{c} M \xrightarrow{F} \mathbb{R}^{q,n} \\ {}^{i_{\mathcal{F}_p}}\uparrow \; \nearrow {}_{F \circ i_{\mathcal{F}_p}} \\ \mathcal{F}_p \end{array},$$

let us drop the index p to make the notation less heavy. As a result, $A_{F \circ i_{\mathcal{F}}}$ and $A_{i_{\mathcal{F}}}$ denote the second fundamental tensors of $F \circ i_{\mathcal{F}}$ and $i_{\mathcal{F}}$ respectively. From equation (2.4), we have that

$$A_{F \circ i_{\mathcal{F}}} = A_F + dF(A_{i_{\mathcal{F}}}).$$

Equation (4.30) implies that $A_{F \circ i_{\mathcal{F}}} = dF(A_{i_{\mathcal{F}}})$. On the other hand, one can write, for vector fields $f_1, f_2 \in \Gamma(T\mathcal{F}_p)$, the second fundamental tensor of the immersion $i_{\mathcal{F}}$ as

$$A_{i_{\mathcal{F}}}(f_1, f_2) = \nabla_{f_1} f_2 - \nabla'_{f_1} f_2, \tag{4.31}$$

where ∇' is the Levi-Civita connection of $\mathcal{F}_p$ (with respect to the first fundamental form of $\mathcal{F}_p$) and we are identifying vectors in $T_p \mathcal{F}$ with vectors in $T_p M$ through $di_{\mathcal{F}}$[8]. From the fact that $\nabla_{f_1} f_2 \in \Gamma(\mathcal{F} M|_{i_{\mathcal{F}}(\mathcal{F}_p)})$ (equation (4.16)) and $\nabla'_{f_1} f_2 \in \Gamma(T\mathcal{F}_p)$ (by definition) it holds that

$$A_{i_{\mathcal{F}}} = 0$$

because $A_{i_{\mathcal{F}}} \in \Gamma(T\mathcal{F}_p^\perp)$. So, $A_{F \circ i_{\mathcal{F}}} = 0$, which means that $\mathcal{F}_p$ is totally geodesic and, analogous to Lemma 4.1.4 for $\mathcal{E}_p$, the geodesics of $\mathcal{F}_p$ are also complete. Writing D for the L.C. connection on $\mathbb{R}^{q,n}$, one has $D_{f_1} f_2 = A_F(f_1, f_2) + \nabla_{f_1} f_2 = \nabla_{f_1} f_2$ so that (the image through $F \circ i_{\mathcal{F}}$ of) the geodesics of $\mathcal{F}_p$ are also geodesics of $\mathbb{R}^{q,n}$ (which are straight lines). Furthermore, $\mathcal{F}_p$ is geodesically complete, thence for any $X \in T_p \mathcal{F}_p$ it holds that $F \circ i_{\mathcal{F}}(p) + dF \circ di_{\mathcal{F}}(X) \in F \circ i_{\mathcal{F}}(\mathcal{F}_p)$ (identifying vectors in $T_{F(p)} \mathbb{R}^{q,n}$ with points in $\mathbb{R}^{q,n}$), so that each connected component of $F \circ i_{\mathcal{F}}(\mathcal{F}_p)$ is an affine $m-r$-dimensional subspace of $\mathbb{R}^{q,n}$ by the linearity of $d(F \circ i_{\mathcal{F}})$, but a leaf of a foliation has only one connected component by definition and it follows that $F \circ i_{\mathcal{F}}(\mathcal{F}_p)$ is an affine $m-r$-dimensional subspace of $\mathbb{R}^{q,n}$.

Let us fix p and prove that $F \circ i_{\mathcal{F}}(\mathcal{F}_{p'})$ is parallel to $F \circ i_{\mathcal{F}}(\mathcal{F}_p)$ for any

---

[8] For more details on this a reference is [dC92].

$p' \in i_{\mathcal{E}}(\mathcal{E}_p)$. For this, let $\gamma : [0,1] \to i_{\mathcal{E}}(\mathcal{E}_p)$ be a smooth curve with $\gamma(0) = p$ and $\gamma(1) = p'$ and let $f_p \in \mathcal{F}M_p$ be a vector. Denote by $f(t)$ the parallel transport of f_p along γ with respect to ∇. From $\nabla P = 0$, it holds that $\nabla_{\dot\gamma}(Pf(t)) = P(\nabla_{\dot\gamma}f(t)) = 0$, which means that $P(f(t))$ is the parallel translation of $P(f_p) = 0$, so that $P(f(t)) = 0$ and $f(t) \in \mathcal{F}M_{i_{\mathcal{E}}\circ\gamma(t)}$ for all $t \in [0,1]$. By writing the second fundamental tensor of F like equation (4.31) we get, using eq. (4.30),

$$D_{\dot\gamma}f(t) = \nabla_{\dot\gamma}f(t) + A(\dot\gamma, f(t)) = 0,$$

where D is the Levi-Civita connection of $\mathbb{R}^{q,n}$. This means that $dF \circ f(t)$ is also the parallel translation of $dF \circ f_p$ in $\mathbb{R}^{q,n}$, which is $dF \circ f(t) = dF \circ f_p$ for all $t \in [0,1]$. This means that $dF \circ di_{\mathcal{F}}(T_p\mathcal{F}_p) \subset dF \circ di_{\mathcal{F}}(T_{p'}\mathcal{F}_p)$. Analogously, $dF \circ di_{\mathcal{F}}(T_{p'}\mathcal{F}_p) \subset dF \circ di_{\mathcal{F}}(T_p\mathcal{F}_p)$. But we already know that the integral leaves of $\mathcal{F}$ are affine subspaces, so that they are equal, up to a translation, to their tangent spaces and thus are parallel. □

All that is left of Theorem 4.1.1 is to show that $F(M)$ is the product $F(\mathcal{E}_q) \times F(\mathcal{F}_q)$, where $q \in M$ minimizes $\|F\|^2$.

Let $q \in M$ be a minimal point of $\|F\|^2$ and $\{f_1, \ldots, f_{m-r}\}$ be an orthonormal basis of $\mathcal{F}M_q$. We define a function $\mathfrak{h} : \mathcal{E}_q \times \mathbb{R}^{m-r} \to F(M)$, given by

$$\mathfrak{h}(p, X) = F(i_{\mathcal{E}}(p)) + X^i dF(f_i) \qquad \forall X = (X^1, \ldots, X^{m-r}) \in \mathbb{R}^{m-r}, p \in \mathcal{E}_q.$$

As all the leaves $\mathcal{F}_{q'}$, $q' \in \mathcal{E}_q$, are parallel, the image of $\mathfrak{h}$ is indeed contained in $F(M)$. Let us consider in $\mathbb{R}^{m-r}$ the canonical metric and in $\mathcal{E}_q \times \mathbb{R}^{m-r}$ the product metric, so that $\mathfrak{h}$ is an isometry because F and $i_{\mathcal{E}}$ are isometries.

$\mathcal{E}_q \times \mathbb{R}^{m-r}$ is geodesically complete (Corollary 6.3.2). We claim that $\mathfrak{h}$ is surjective. To see this, take $(p, X) \in \mathcal{E}_q \times \mathbb{R}^{m-r}$, $y := \mathfrak{h}(p, X) \in F(M)$, and $z \in F(M)$. Let $y' \in M$ and $z' \in M$ be such that $F(y') = y$ and $F(z') = z$. From the fact that M is geodesically complete (hypothesis), there is a vector $Y \in T_{y'}M$ such that $\exp(Y) = z'$ (by the Theorem of Hopf and Rinow, Thr. 6.3.1). Then decompose $Y = Y_1 + Y_2$ with $Y_1 \in T_p\mathcal{E}_q$ and $Y_2 = Y_2^l f_l(p) \in \mathcal{F}M_p$. Now denote

$Y_{20} := (Y_2^1, \ldots, Y_2^{m-r})$, then for the exponential in $\mathcal{E}_q \times \mathbb{R}^{m-r}$ it holds that

$$\mathfrak{h}(\exp(Y_1, Y_{20})) = \exp(d\mathfrak{h}(Y_1, Y_{20})) = \exp(dF \circ di_{\mathcal{E}}(Y_1) + dF(Y_2))$$
$$= F(\exp(di_{\mathcal{E}}(Y_1) + Y_2) = F(\exp(Y)) = z,$$

where we understand $F(M)$ locally as a manifold (isometric to M and with the same dimension) and thence define the exponential there locally, so that, by the compactness of the domain of the geodesic segment connecting y' and z', the exponential is well defined. This proves that $z \in \mathfrak{h}(\mathcal{E}_q \times \mathbb{R}^{m-r})$.

Then $F(M)$ is the product of an affine space with a minimal surface of the hyperquadric $\mathcal{H}^{n-1}(r)$ with $\|\vec{H}\|^2 = r$. □

Remark 4.1.7. The induced (from $\mathbb{R}^{q,n}$) inner product on the affine space has to be positive definite, because F is spacelike.

4.2 The Second Case

As a remark to the notation: Throughout the proof of the next Theorem, we identify vectors, vectorfields and some curves in different manifolds through the immersions among then without explicit mention several times in order to avoid the notation becoming too heavy.

Let us now consider the case where $\frac{\nabla\|\vec{H}\|}{\|\nabla\|\vec{H}\|\|}$ is the only eigenvector associated with a non-zero eigenvalue of P.

Theorem 4.2.1. *Let M be a smooth manifold and $F : M \to \mathbb{R}^{q,n}$ be a mainly positive, spacelike, shrinking self-similar solution of the mean curvature flow with bounded geometry such that $F(M)$ is unbounded. Beyond that, let F satisfy the conditions: $\|\vec{H}\|^2(p) \neq 0$ for all $p \in M$ and the principal normal is parallel in the normal bundle ($\nabla^\perp \nu \equiv 0$). If $\nabla\|\vec{H}\|(p) \neq 0$ for some $p \in M$, then*

$$F(M) = \Gamma \times \mathbb{R}^{m-1}, \tag{4.32}$$

where Γ is a rescaling of an Abresch & Langer curve in a spacelike plane and R^{m-1} is an $m-1$ dimensional spacelike affine space in $\mathbb{R}^{q,n}$.

Proof. Let $p \in M$ be a point with $\nabla\|\vec{H}\|(p) \neq 0$ and $\{e_i\}_{i=1,\ldots,m}$ an orthonormal basis of T_pM made by the eigenvectors of P with $e_1 = \frac{\nabla\|\vec{H}\|}{\|\nabla\|\vec{H}\|\|}(p)$. We have, just as in Lemma 4.0.28, that the only nonzero element of the matrix that represents P in this base is the element in the first row and first column. But $\operatorname{tr}(P) = \Sigma_i \langle \vec{H}, A(e_i, e_i) \rangle = \|\vec{H}\|^2$, so this first element of the matrix has to be $\|\vec{H}\|^2$ and the eigenvalue associated with $\frac{\nabla \|\vec{H}\|}{\|\nabla \|\vec{H}\|\|}$ is $\|\vec{H}\|^2$:

$$P^i_j \frac{\nabla^j \|\vec{H}\|}{\|\nabla \|\vec{H}\|\|} = \|\vec{H}\|^2 \frac{\nabla^i \|\vec{H}\|}{\|\nabla \|\vec{H}\|\|} \tag{4.33}$$

and

$$\|P\|^2 = \|\vec{H}\|^4$$

at this point, but this equation holds all over M because equation (4.5) together with the fact that $\nabla_i \|\vec{H}\| \frac{P_{jk}}{\|\vec{H}\|} - \|\vec{H}\| \nabla_i \left(\frac{P_{jk}}{\|\vec{H}\|} \right) = 0$ (Lemma 4.0.27) implies that

$$\int_M \rho \|\vec{H}\|^2 \left\| \nabla \left(\frac{\|P\|^2}{\|\vec{H}\|^4} \right) \right\|^2 d\mu = 0$$

and

$$\nabla \left(\frac{\|P\|^2}{\|\vec{H}\|^4} \right) = 0$$

because M is spacelike and $\|\vec{H}\|^2 \neq 0$.

Remark 4.2.2. Let us choose Riemannian normal coordinates on a neighborhood of p such that $\frac{\partial}{\partial x^i}(p) = e_i$, then it holds in p: $P_{ij} = \|\vec{H}\|^2 \delta_{1i} \delta_{1j}$ and $g_{ij} = \delta_{ij}$, thus it follows that $P^k_i A_{kj} = 0$ if $i \neq 1$ and, from Lemma 3.2.3, item (3), that $P^k_i A_{kj} = 0$ if $j \neq 1$ so that $P^k_1 A^1_k = P^k_i A^i_k$, which is in the direction of $\nu/\|\vec{H}\|$ from Lemma 3.2.3 (item 1), thence the component of $P^k_i A_{kj}$ in $F^*(\mathbb{R}^{q,n})$ has the same direction as $\frac{\nu}{\|\vec{H}\|}$ for all $i, j \in 1, \ldots, m$ and

$$P^k_i A_{kj} = P^k_i \langle \nu, A_{kj} \rangle \nu = \frac{1}{\|\vec{H}\|} P^k_i P_{kj} \nu = \|\vec{H}\|^3 \delta_{1i} \delta^k_1 \delta_{1k} \delta_{1j} \nu = \|\vec{H}\| P_{ij} \nu,$$

i. e. $P * A = \|\vec{H}\| P \otimes \nu$, but this can be done for any $p \in M$ with $\nabla \|\vec{H}\|(p) \neq 0$, so that this tensor equality holds in any region of M that satisfies $\nabla \|\vec{H}\|(p) \neq 0$, and this is written in (any) local coordinates as

$$P^k_i A_{kj} = \|\vec{H}\| P_{ij} \nu. \tag{4.34}$$

Let us now define
$$\mathring{M} = \{p \in M : \nabla \|\vec{H}\| \neq 0\}.$$

Note that $\mathring{M}$ is the inverse image of the set $(-\infty, 0) \cup (0, \infty)$ trough the function $\mathfrak{h}: M \to \mathbb{R}$, $\mathfrak{h}(p) := \|\nabla\|\vec{H}\|\|^2$, which is continuous, so that $\mathring{M}$ is open, and thus a submanifold of M (possibly incomplete). Let $U \subset M$ be a connected component of $\mathring{M}$. We consider over U the distributions $\mathcal{E}U$ and $\mathcal{F}U$ given on each point by

$$\mathcal{E}U_p := \{V \in T_p U : PV = \|\vec{H}\|^2 V\}, \tag{4.35}$$

$$\mathcal{F}U_p := \{V \in T_p U : PV = 0\}. \tag{4.36}$$

In order to investigate these distributions we need further information about the second funcamental tensor. For this purpose we define the tensor $\mathring{A} := A - \frac{1}{\|\vec{H}\|} P \otimes \nu$, which is written in local coordinates

$$\mathring{A}_{ij} = A_{ij} - \frac{1}{\|\vec{H}\|} P_{ij} \nu.$$

Lemma 4.2.3. *Under the hypothesis of Theorem 4.2.1, it holds*

$$A_{ij} = \frac{1}{\|\vec{H}\|} P_{ij} \nu.$$

Proof. From eq. (3.7) we calculate

$$\|\vec{H}\| \nabla_i \|\vec{H}\| = \langle \nabla_i^\perp \vec{H}, \vec{H} \rangle = \theta^k P_{ik}, \tag{4.37}$$

this equation together with the fact that $\frac{\nabla \|\vec{H}\|}{\|\nabla\|\vec{H}\|\|}$ is an eigenvector of P (associated with the eigenvalue $\|\vec{H}\|^2$) implies

$$\theta(\nabla\|\vec{H}\|) = \frac{\theta^k}{\|\vec{H}\|^2} P_k^l(\nabla_l\|\vec{H}\|) = \frac{\|\vec{H}\|\nabla^l\|\vec{H}\|\nabla_l\|\vec{H}\|}{\|\vec{H}\|^2} = \frac{\|\nabla\|\vec{H}\|\|^2}{\|\vec{H}\|}. \tag{4.38}$$

In order to attain $\mathring{A}_{ij} = 0$ we will consider, as in the first case, the integral curves of a certain vectorfield: The projection of F in $\mathcal{F}U$, that is

$$\mathring{\theta}_i = \theta_i - \frac{\theta(\nabla^k \|\vec{H}\|)}{\|\nabla_k\|\vec{H}\|\|^2} \nabla_i\|\vec{H}\| = \theta_i - \frac{1}{\|\vec{H}\|} \nabla_i\|\vec{H}\|.$$

For $\mathring{A}$ one calculates the following:

$$\mathring{\partial}^k \mathring{A}_{ki} = \left(\theta^k - \frac{1}{\|\vec{H}\|}\nabla^k\|\vec{H}\|\right)\left(A_{ki} - \frac{1}{\|\vec{H}\|}P_{ki}\nu\right)$$

$$\mathring{\partial}^k \mathring{A}_{ki} = \theta^k A_{ki} - \nabla_i\|\vec{H}\|\nu - \frac{1}{\|\vec{H}\|}\nabla^k\|\vec{H}\|A_{ki} + \frac{1}{\|\vec{H}\|^2}\nabla^k\|\vec{H}\|P_{ki}\nu$$

$$= \theta^k A_{ki} - \frac{1}{\|\vec{H}\|^3}\nabla^j\|\vec{H}\|P_j^k A_{ki} = \theta^k A_{ki} - \frac{1}{\|\vec{H}\|^2}\nabla^j\|\vec{H}\|P_{ij}\nu$$

$$= \theta^k A_{ki} - \nabla_i\|\vec{H}\|\nu = \theta^k A_{ki} - \nabla_i^\perp \vec{H} = 0,$$

where we used (4.33), (4.34) and (4.37) (and eq. (3.7) in the last step). From this, the following can be deduced

$$0 = \nabla_l^\perp(\mathring{\partial}^k \mathring{A}_{ki})$$

$$= \left(\delta_l^k - P_l^k + \frac{1}{\|\vec{H}\|^2}\nabla_l\|\vec{H}\|\nabla^k\|\vec{H}\| - \frac{1}{\|\vec{H}\|}\nabla_l\nabla^k\|\vec{H}\|\right)\mathring{A}_{ki} + \mathring{\partial}^k \nabla_l^\perp \mathring{A}_{ki}$$

$$0 = \mathring{A}_{li} - \frac{1}{\|\vec{H}\|}\nabla_l\nabla^k\|\vec{H}\|\mathring{A}_{ki} + \mathring{\partial}^k \nabla_l^\perp \mathring{A}_{ki}, \qquad (4.39)$$

because equations (4.33) and (4.34) with $P_l^k P_{ki} = \|\vec{H}\|^2 P_{li}$ imply that $P_l^k \mathring{A}_{ki} = 0$ and equations (4.33), (4.34) and (4.37) imply

$$\nabla^k\|\vec{H}\|\mathring{A}_{ki} = \nabla^k\|\vec{H}\|A_{ki} - \nabla^k\|\vec{H}\|\frac{1}{\|\vec{H}\|}P_{ki}\nu$$

$$= \frac{1}{\|\vec{H}\|}\theta^l P_l^k A_{ki} - \|\vec{H}\|\nabla_i\|\vec{H}\|\nu$$

$$= \theta^l P_{li}\nu - \|\vec{H}\|\nabla_i\|\vec{H}\|\nu = \|\vec{H}\|\nabla_i\|\vec{H}\|\nu - \|\vec{H}\|\nabla_i\|\vec{H}\|\nu = 0.$$

On the other hand, take a point $p \in U$, Riemannian normal coordinates in a neighborhood of this point such that $\frac{\partial}{\partial x^1} = \frac{\nabla\|\vec{H}\|}{\|\nabla\|\vec{H}\|\|}$ and calculate for $i \in 1,\ldots,n$ and $j \in 2,\ldots,n$,

$$\nabla_i \nabla_j \|\vec{H}\| = \frac{\partial}{\partial x^i}\left\langle \nabla\|\vec{H}\|, \frac{\partial}{\partial x^j}\right\rangle - \left\langle \nabla\|\vec{H}\|, \nabla_{\frac{\partial}{\partial x^i}}\frac{\partial}{\partial x^j}\right\rangle = 0$$

because $\frac{\partial}{\partial x^j} \in \mathcal{F}U_p$ and $\nabla_{\frac{\partial}{\partial x^i}}\frac{\partial}{\partial x^j}(p) = 0$. But $\nabla_i\nabla_j\|\vec{H}\|$ is symmetric (one can see this writing it in local coordinates). So that $\nabla_i \nabla_j \|\vec{H}\|$ is nonzero only if $i = j = 1$, but the trace of it is the Laplacian, so that using $P_{ij} = \|\vec{H}\|^2 \delta_{1i}\delta_{1j}$ we get

$$\nabla_i \nabla_j \|\vec{H}\| = \frac{\triangle \|\vec{H}\|}{\|\vec{H}\|^2} P_{ij}, \qquad (4.40)$$

but this equation is tensorial [9], so that this holds for any choice of coordinates. And, as we could do the same for every point $p \in U$, this holds in the whole $\mathring{M}$. So that, with eq. (4.34),

$$\begin{aligned}\nabla_l \nabla^k \|\vec{H}\| \mathring{A}_{ki} &= \frac{\triangle \|\vec{H}\|}{\|H\|^2} P_l^k \mathring{A}_{ki} \\ &= \frac{\triangle \|\vec{H}\|}{\|H\|^2} P_l^k \left(A_{ki} - \frac{1}{\|\vec{H}\|} P_{ki} \nu \right) \\ &= \frac{\triangle \|\vec{H}\|}{\|H\|^2} \left(\|\vec{H}\| P_{li} \nu - \|\vec{H}\| P_{li} \nu \right) = 0 \end{aligned}$$

and equation (4.39) turns out to be

$$\mathring{\theta}^k \nabla_l^\perp \mathring{A}_{ki} = -\mathring{A}_{li} \qquad (4.41)$$

Finally, we calculate using eq. (4.37)

$$\begin{aligned}\nabla_i \|\mathring{\theta}\|^2 &= 2\nabla_i \mathring{\theta}_l \mathring{\theta}^l \\ &= 2\left(\theta_i - \theta^l P_{li} + \theta^l \frac{\nabla_i \|\vec{H}\| \nabla_l \|\vec{H}\|}{\|\vec{H}\|^2} - \frac{\triangle \|\vec{H}\|}{\|\vec{H}\|^3} P_{il} \theta^l - \frac{\nabla_i \|\vec{H}\|}{\|\vec{H}\|} \right. \\ &\quad \left. + \frac{\nabla^l \|\vec{H}\|}{\|\vec{H}\|} P_{li} - \frac{\|\nabla \|\vec{H}\|\|^2}{\|\vec{H}\|^3} \nabla_i \|\vec{H}\| + \frac{\triangle \|\vec{H}\|}{\|\vec{H}\|^4} P_{il} \nabla^l \|\vec{H}\| \right) \\ &= 2\left(\theta_i + \theta^l \frac{\nabla_i \|\vec{H}\| \nabla_l \|\vec{H}\|}{\|\vec{H}\|^2} - \frac{\nabla_i \|\vec{H}\|}{\|\vec{H}\|} - \frac{\|\nabla \|\vec{H}\|\|^2}{\|\vec{H}\|^3} \nabla_i \|\vec{H}\| \right),\end{aligned}$$

---

[9] Eq. (4.40) can be written as $\nabla d \|\vec{H}\| = \frac{\triangle \|\vec{H}\|}{\|\vec{H}\|^2} P$.

but using eq. (4.37)

$$\mathring{\theta}^i \nabla_i \|\vec{H}\| = \langle F, F^i\rangle \nabla_i\|\vec{H}\| - \frac{1}{\|\vec{H}\|^2}\langle F, F_k\rangle P^{ki}\nabla_i\|\vec{H}\|$$
$$= \langle F, F^i\rangle \nabla_i\|\vec{H}\| - \langle F, F^k\rangle \nabla_k\|\vec{H}\| = 0.$$

which was already expected since $\mathring{\theta}$ is the pointwise projection of F onto $\mathcal{F}U$ and $\frac{\nabla\|\vec{H}\|}{\|\nabla\|\vec{H}\|\|} \in \Gamma(\mathcal{E}U)$, so that it holds

$$\mathring{\theta}(\nabla\|\mathring{\theta}\|^2) = 2\mathring{\theta}^i\theta_i = 2\mathring{\theta}^i\left(\theta_i - \frac{\nabla_i\|\vec{H}\|}{\|\vec{H}\|}\right) = 2\|\mathring{\theta}\|^2. \qquad (4.42)$$

Now we can prove that $\mathring{A}_{ij} = 0$. If $\mathring{\theta} = 0$ at some point $p \in U$, then equation (4.41) implies $\mathring{A}_{ij} = 0$ and $A_{ij} = \frac{1}{\|\vec{H}\|}P_{ij}\nu$ at this point. So, without loss of generality, we can consider only the points $q \in U$ with $\mathring{\theta}(q) \neq 0$. Take one of these fixed and consider the maximal integral curve $\gamma : (-a, b) \to M$ of $\mathring{\theta}^k$, with $\gamma(0) = q$ and $a, b > 0 \in \mathbb{R}$. Although we calculated (4.42) in a connected component of U, it is clear that this holds in the whole $\mathring{M}$ and, by continuity, in $\overline{\mathring{M}}$. These integral curves can go outside $\mathring{M}$, but in open sets of $M \setminus \mathring{M}$ the equations, up to eq. (4.23), of the first case hold because $M \setminus \mathring{M}$ is given by the points with $\nabla\|\vec{H}\| = 0$ and (until this equation) everything is calculated locally in the first case. This way, in an open set of $M \setminus \mathring{M}$, it holds $\mathring{\theta} = \theta$ and this implies, from eq. (4.23), that equation (4.42) holds globally in M. Along this curve we define the function

$$f(s) := \|\mathring{\theta}\|^2(\gamma(s)),$$

and get

$$\frac{d}{ds}f = \nabla_{\dot\gamma}\|\mathring{\theta}\|^2 = \mathring{\theta}^k\nabla_k\|\mathring{\theta}\|^2 = 2\|\mathring{\theta}\|^2$$

using equation (4.42) (and $\mathring{\theta} = \theta$ in the open sets of $M \setminus \mathring{M}$). But this equation has a unique solution with $f(0) = \|\mathring{\theta}\|^2(q)$

$$\|\mathring{\theta}\|^2(\gamma(s)) = \|\mathring{\theta}\|^2(q)e^{2s} > 0,$$

this means these integral curves do not cross any singular point and thence the maximal integral curve is defined in all $\mathbb{R}$ ($a = \infty, b = \infty$) and it is not closed (because of injectivity of e^{2t}). Over this same curve we define other functions $\tilde{f}_\pm : \mathbb{R} \to \mathbb{R}$,

$$\tilde{f}_\pm(s) := (\|\mathring{A}_\pm\|^2)(\gamma(s))$$

and for the derivative of this function, using equation (4.41), we get in $\overline{\mathring{M}}$

$$\begin{aligned}\frac{d\tilde{f}}{ds} &= \dot{\gamma}(\|\mathring{A}_\pm\|^2) = 2\mathring{\theta}^k\left\langle \nabla_k^\perp(\mathring{A}_\pm)_{ij}, \mathring{A}_\pm^{ij}\right\rangle \\ &= -2\left\langle (\mathring{A}_\pm)_{ij}, \mathring{A}_\pm^{ij}\right\rangle = -2\tilde{f}.\end{aligned}$$

We still have to check if this holds in open sets of $M \setminus \mathring{M}$. In open sets of $M \setminus \mathring{M}$, $P = P^2$ (eq. (4.11)) implies that the only non-zero eigenvalue of P is 1, then $\nabla P = 0$[10] together with $\|P\|^2 = r$, where r is the multiplicity of the eigenvalue 1, implies that r is constant, but M is connected and, in $\overline{\mathring{M}}$, the tensor P has only one non-zero eigenvalue and it has multiplicity 1, then P has only one non-zero eigenvalue and it has multiplicity 1 also in open sets of $M \setminus \mathring{M}$. Therefore, as in Remark 4.2.2, we get that $P_{ik}A_j^k$ is in the direction of $\frac{\vec{H}}{\|\vec{H}\|}$ and

$$\begin{aligned}\|\mathring{A}_\pm\|^2 &= \left\|A_\pm - \frac{1}{\|\vec{H}\|}P \otimes \nu_\pm\right\|^2 = \|A_\pm\|^2 - \frac{2}{\|\vec{H}\|}\langle A_\pm^{ij}, P_{ij}\nu_\pm\rangle + \frac{1}{\|\vec{H}\|^4}\|P\|^2\|\vec{H}_\pm\|^2 \\ &= \|A_\pm\|^2 - \frac{2}{\|\vec{H}\|^4}\langle P^{ij}P_{ij}\vec{H}_\pm, \vec{H}_\pm\rangle + \frac{1}{\|\vec{H}\|^4}\|P\|^2\|\vec{H}_\pm\|^2 \\ \|\mathring{A}_\pm\|^2 &= \|A_\pm\|^2 - \frac{1}{\|\vec{H}\|^4}\|P\|^2\|\vec{H}_\pm\|^2,\end{aligned} \qquad (4.43)$$

and

$$\begin{aligned}\|P * A_\pm\|^2 &= \langle P_{ik}(A_\pm)_j^k, P_l^i A_\pm^{lj}\rangle = \left\langle P_{ik}\left\langle A_j^k, \frac{\vec{H}}{\|\vec{H}\|}\right\rangle\frac{\vec{H}_\pm}{\|\vec{H}\|}, P_l^i\left\langle A^{lj}, \frac{\vec{H}}{\|\vec{H}\|}\right\rangle\frac{\vec{H}_\pm}{\|\vec{H}\|}\right\rangle \\ &= P_{ik}P_j^k P_l^i P^{lj}\frac{1}{\|\vec{H}\|^4}\|\vec{H}_\pm\|^2 = \frac{\|P\|^2}{\|\vec{H}\|^4}\|\vec{H}_\pm\|^2,\end{aligned}$$

---

[10]This is at the beginning of the first case.

using $P_{ij} = P_{ik}P_j^k$ (eq. (4.11)), thence, by eq. 4.43, it holds

$$\|\mathring{A}_\pm\|^2 = \|A_\pm\|^2 - \frac{1}{\|\vec{H}\|^4}\|P\|^2\|\vec{H}_\pm\|^2 = \|A_\pm\|^2 - \|P*A_\pm\|^2 \qquad (4.44)$$

and equation (4.22) implies that $\frac{d\tilde{f}}{ds} = -2\tilde{f}$ holds in the whole manifold M. This O.D.E. has a unique solution with $\tilde{f}(0) = \|\mathring{A}\|^2(q)$:

$$(\|\mathring{A}\|^2)(\gamma(s)) = (\|\mathring{A}\|^2)(q)e^{-2s}.$$

If $(\|\mathring{A}_\pm\|^2)(q) \neq 0$ then $(\|\mathring{A}_\pm\|^2)(\gamma(s)) \to \pm\infty$ as $s \to -\infty$ and this contradicts the boundedness of the second fundamental tensor (by the definition of bounded geometry). So that $\|\mathring{A}_\pm\|^2 = 0$ and

$$A_{ij} = \frac{1}{\|\vec{H}\|}P_{ij}\nu \qquad (4.45)$$

holds in $\mathring{M}$ as we wanted to show. $\square$

We prove now that the distributions $\mathcal{E}U$ and $\mathcal{F}U$ are involutive.

Lemma 4.2.4. *Under the hypothesis of Theorem 4.2.1 the distributions $\mathcal{E}U$ and $\mathcal{F}U$ are involutive.*

Proof. First of all recall that $\mathcal{E}U$ is one dimensional and is spanned by $\nabla\|\vec{H}\|/\|\nabla\|\vec{H}\|\|(p)$ at any $p \in U$. By equation 4.40 it holds that $\nabla_i\nabla_j\|\vec{H}\| = \nabla_j\nabla_i\|\vec{H}\|$ and it follows

$$\nabla_l\|\nabla\|\vec{H}\|\|^2 = 2\nabla_l\nabla^k\|\vec{H}\|\nabla_k\|\vec{H}\| = 2\nabla^k\nabla_l\|\vec{H}\|\nabla_k\|\vec{H}\|$$
$$\implies \nabla\|\nabla\|\vec{H}\|\|^2 = 2\nabla_{\nabla\|\vec{H}\|}\nabla\|\vec{H}\|.$$

Then let $p \in U$ be a point and $X_p \in T_pU$ be normal to $\nabla\|\vec{H}\|/\|\nabla\|\vec{H}\|\|(p)$; beyond this, let $X \in \Gamma(TU|_V)$ be the parallel transport of X_p over all geodesics through p in a small neighborhood V of p. Then at p

$$\langle \nabla\|\nabla\|\vec{H}\|\|^2, X\rangle = 2\langle \nabla_{\nabla\|\vec{H}\|}\nabla\|\vec{H}\|, X\rangle$$
$$= 2\nabla\|\vec{H}\|\langle \nabla\|\vec{H}\|, X\rangle - 2\langle \nabla\|\vec{H}\|, \nabla_{\nabla\|\vec{H}\|}X\rangle = 0,$$

so that from $\nabla \|\nabla \|\vec{H}\|\|^2 = 2(\|\nabla \|\vec{H}\|\|)\nabla \|\nabla \|\vec{H}\|\|$ it holds that $(\|\nabla \|\vec{H}\|\|)\nabla \|\nabla \|\vec{H}\|\|(p) \in \mathcal{E}U_p$. But this holds for any p in U so that $(\|\nabla \|\vec{H}\|\|)\nabla \|\nabla \|\vec{H}\|\| \in \Gamma(\mathcal{E}U)$. Therefore

$$\frac{\nabla^k\|\vec{H}\|}{\|\nabla\|\vec{H}\|\|}\nabla_k\left(\frac{\nabla_l\|\vec{H}\|}{\|\nabla\|\vec{H}\|\|}\right) = \frac{\nabla^k\|\vec{H}\|}{\|\nabla\|\vec{H}\|\|}\frac{\nabla_k\nabla_l\|\vec{H}\|}{\|\nabla\|\vec{H}\|\|} - \frac{\nabla^k\|\vec{H}\|}{\|\nabla\|\vec{H}\|\|}\nabla_k\|\nabla\|\vec{H}\|\|\frac{\nabla_l\|\vec{H}\|}{\|\nabla\|\vec{H}\|\|^2}$$

$$= \frac{1}{2}\nabla_l\left(\frac{\nabla^k\|\vec{H}\|}{\|\nabla\|\vec{H}\|\|}\frac{\nabla_k\|\vec{H}\|}{\|\nabla\|\vec{H}\|\|}\right) - \frac{\nabla^k\|\vec{H}\|}{\|\nabla\|\vec{H}\|\|}\nabla_k\|\nabla\|\vec{H}\|\|\frac{\nabla_l\|\vec{H}\|}{\|\nabla\|\vec{H}\|\|^2}$$
$$+ \frac{\nabla^k\|\vec{H}\|}{\|\nabla\|\vec{H}\|\|}\nabla_l\|\nabla\|\vec{H}\|\|\frac{\nabla_k\|\vec{H}\|}{\|\nabla\|\vec{H}\|\|^2}$$
$$= \frac{\nabla_l\|\nabla\|\vec{H}\|\|}{\|\nabla\|\vec{H}\|\|} - \frac{\nabla^k\|\vec{H}\|}{\|\nabla\|\vec{H}\|\|}\nabla_k\|\nabla\|\vec{H}\|\|\frac{\nabla_l\|\vec{H}\|}{\|\nabla\|\vec{H}\|\|^2}$$

$$\nabla_{\frac{\nabla\|\vec{H}\|}{\|\nabla\|\vec{H}\|\|}}\frac{\nabla\|\vec{H}\|}{\|\nabla\|\vec{H}\|\|} = \frac{\nabla\|\nabla\|\vec{H}\|\|}{\|\nabla\|\vec{H}\|\|} - \left(\frac{\nabla^k\|\vec{H}\|}{\|\nabla\|\vec{H}\|\|^3}\nabla_k\|\nabla\|\vec{H}\|\|\right)\nabla\|\vec{H}\| \in \Gamma(\mathcal{E}U).$$

Then, as any $X, Y \in \Gamma(\mathcal{E}U)$ are of the form $X = x\frac{\nabla\|\vec{H}\|}{\|\nabla\|\vec{H}\|\|}$ and $Y = y\frac{\nabla\|\vec{H}\|}{\|\nabla\|\vec{H}\|\|}$ for some $x, y \in C^\infty(U)$, it holds that

$$\nabla_X Y = x\left(\frac{\nabla\|\vec{H}\|}{\|\nabla\|\vec{H}\|\|}(y)\right)\frac{\nabla\|\vec{H}\|}{\|\nabla\|\vec{H}\|\|} + xy\nabla_{\frac{\nabla\|\vec{H}\|}{\|\nabla\|\vec{H}\|\|}}\frac{\nabla\|\vec{H}\|}{\|\nabla\|\vec{H}\|\|} \in \Gamma(\mathcal{E}U). \quad (4.46)$$

But this could be done for any $p \in U$, so that, just as in the first case, $\mathcal{E}U$ is in particular involutive. So, by the Theorem of Frobenius, there is a foliation, whose tangent spaces of the leaves are given by this distribution. The leaves are immersed in M and, again as in the first case, they are totally geodesic (analogous to eq. (4.25) and eq. (4.26)). This means, in particular, for any $p \in U$, that a geodesic of the one dimensional leaf $(\mathcal{E}_p)$ that goes through p is also a geodesic of U.

Let $p \in U$ be a fixed point, and take Riemannian normal coordinates on a neighborhood of p such that $\frac{\partial}{\partial x^1}(p) = \frac{\nabla\|\vec{H}\|}{\|\nabla\|\vec{H}\|\|}$. This way the tensor P is written, in these coordinates and at this point p, as $P_{ij} = \|\vec{H}\|^2 \delta_{1i}\delta_{1j}$. So, for

$V, W \in \Gamma(\mathcal{F}U)$, using the fact that $\mathcal{F}U_p \perp \mathcal{E}U_p$ and $\frac{\nabla \|\vec{H}\|}{\|\nabla \|\vec{H}\|\|} \in \Gamma(\mathcal{E}U)$, we get

$$0 = \langle \nabla \|\vec{H}\|, V \rangle = \nabla_V \|\vec{H}\|,$$

from this, remembering $\Gamma_{ij}^k(p) = 0$, follows

$$\nabla_V P = \nabla_V \|\vec{H}\|^2 \delta_{1i} \delta_{1j} = 0$$

and

$$P(\nabla_V W) = \nabla_V (PW) = 0.$$

This means
$$\nabla_V W \in \Gamma(\mathcal{F}U) \ \forall \ V, W \in \Gamma(\mathcal{F}U). \tag{4.47}$$

As this holds for any $p \in U$ and the final expressions do not depend on local coordinates this holds in the whole $\overset{\circ}{M}$ and $\mathcal{F}U$ is involutive. $\square$

Remark 4.2.5. It follows then, by the Theorem of Frobenius, that the distribution $\mathcal{F}U$ gives rise to a foliation, whose leaf through a point $p \in M$ is written $\mathcal{F}_p$ and the distribution $\mathcal{E}U$ gives rise to a foliation, whose leaf through a point $p \in M$ is written $\mathcal{E}_p$. Beyond this, the leaves are immersed in M with immersions $i_\mathcal{E}$ and $i_\mathcal{F}$.

Now we can see what these integral leaves are.

Lemma 4.2.6. *Under the hypothesis of Theorem 4.2.1, it holds that $F \circ i_\mathcal{F}(\mathcal{F}_p)$ is an affine space in $\mathbb{R}^{q,n}$ for all $p \in U$. Beyond that, if $q \in i_\mathcal{E}(\mathcal{E}_p)$, then $F \circ i_\mathcal{F}(\mathcal{F}_p)$ and $F \circ i_\mathcal{F}(\mathcal{F}_q)$ are parallel.*

Proof. Let $p \in U$ be a point and $\mathcal{F}_p$ be the leaf of the distribution $\mathcal{F}U$ that goes through this point. First, we show this leaf is an affine subspaces of $\mathbb{R}^{q,n}$. Let $q \in i_\mathcal{F}(\mathcal{F}_p)$ be an arbitrary point, $f \in \mathcal{F}U_q$ and $X \in T_q M$ be vectors, then equation (4.45) implies

$$A(f, X) = X^j f^i A_{ij} = X^j f^i \frac{1}{\|\vec{H}\|} P_{ij} \nu = 0 \tag{4.48}$$

because $\mathcal{F}M$ is formed by the vectors in the null space of P.

It is clear that the natural inclusion $i_{\mathcal{F}_p} : \mathcal{F}_p \to U \subset M$ is an immersion[11], so that $F \circ i_{\mathcal{F}_p} : \mathcal{F}_p \to \mathbb{R}^{q,n}$ is also an immersion and we can consider the second fundamental tensors of them:

$$\begin{array}{c} M \xrightarrow{F} \mathbb{R}^{q,n} \\ i_{\mathcal{F}_p} \uparrow \nearrow_{F \circ i_{\mathcal{F}_p}} \\ \mathcal{F}_p \end{array}$$

let us drop the index p to make the notation less heavy. As a result, $A_{F \circ i_{\mathcal{F}}}$ and $A_{i_{\mathcal{F}}}$ denote the second fundamental tensors of $F \circ i_{\mathcal{F}}$ and $i_{\mathcal{F}}$ respectively. From equation (2.4) we know that

$$A_{F \circ i_{\mathcal{F}}} = A_F + dF(A_{i_{\mathcal{F}}}).$$

Equation (4.48) implies that A_F vanishes if applied to vectors of the distribution $\mathcal{F}U$. On the other hand one can write, for vector fields $f_1, f_2 \in \Gamma(T\mathcal{F}_p)$, the second fundamental tensor of the immersion $i_{\mathcal{F}}$ as

$$A_{i_{\mathcal{F}}}(f_1, f_2) = \nabla_{f_1} f_2 - \nabla'_{f_1} f_2, \tag{4.49}$$

where ∇' is the Levi-Civita connection of $\mathcal{F}_p$ (with respect to the first fundamental form of $\mathcal{F}_p$) and we identify vectors in $T_q \mathcal{F}_p$ with vectors in $\mathcal{F}M_q$ through $di_{\mathcal{F}}$. From the fact that $\nabla_{f_1} f_2 \in \Gamma(\mathcal{F}U)$ (by equation (4.47)) and $di_{\mathcal{F}}(\nabla'_{f_1} f_2)(x) \in \mathcal{F}U_{i_{\mathcal{F}}(x)}$ (by definition) it holds that

$$A_{i_{\mathcal{F}}} = 0, \quad \text{thus} \quad \nabla_{f_1} f_2 = \nabla'_{f_1} f_2, \tag{4.50}$$

because $A_{i_{\mathcal{F}}}(f_1, f_2) \in \Gamma(T\mathcal{F}_p^\perp)$. As a result,

$$A_{F \circ i_{\mathcal{F}}} = 0,$$

i. e. $F \circ i_{\mathcal{F}}$ is totally geodesic and

$$D_{f_1} f_2 = A(f_1, f_2) + \nabla_{f_1} f_2 = \nabla_{f_1} f_2 = \nabla'_{f_1} f_2 \tag{4.51}$$

---

[11] One can see this remembering that the charts of the leaves are induced by the charts of M.

implies that the (image through $F\circ i_{\mathcal{F}}$ of the) geodesics of $\mathcal{F}_p$ are also geodesics of $\mathbb{R}^{q,n}$, which are straight lines, but U is not geodesically complete, so that the (image through $F\circ i_{\mathcal{F}}$ of the) geodesics of $\mathcal{F}_p$ could only be some intervals of these lines.

We prove now that $\mathcal{F}_p$ is geodesically complete. Let $\delta: (-a,b) \to \mathcal{F}_p$, $a,b > 0$, be a maximally extended geodesic of $\mathcal{F}_p$ and $\gamma := i_{\mathcal{F}} \circ \delta$, as M is geodesically complete, γ can be extended $\gamma : \mathbb{R} \to M$. Just as in Note 4.1.5 the equations $A_{i_{\mathcal{F}}} = 0$, $P(\dot\gamma(t)) = 0 \,\forall t \in (-a,b)$ and the geodesic completeness of M imply that if $a \neq \infty$ or $b \neq \infty$, then δ could be further extended (which contradicts the fact that δ is maximally extended) as long as $\gamma(t) \in U$.

We claim that $\gamma(t) \in U$ for all $t \in [-a,b]$. To prove this, we need to show that $\nabla \|\vec{H}\|(\gamma(t)) \neq 0$ for $t = -a$ and $t = b$. By equation (4.48) it holds for every $t \in (-a,b)$

$$\nabla_{\dot\gamma} \nabla \|\vec{H}\| = 0,$$

so that $\nabla \|\vec{H}\|(\gamma(t))$ is the parallel transport of $\nabla \|\vec{H}\|(\gamma(0))$. Beyond this, eq. (4.48) also implies

$$D_{\dot\gamma} \nabla \|\vec{H}\| = \nabla_{\dot\gamma} \nabla \|\vec{H}\|$$

so that $dF(\nabla \|\vec{H}\|(\gamma(t)))$ is the parallel transport of $dF(\nabla \|\vec{H}\|(\gamma(0)))$ over the line[12] in $\mathbb{R}^{q,n}$ defined by $dF(\dot\gamma(0))$ and $F(\dot\gamma(0))$. This means that $dF(\nabla \|\vec{H}\|)(\gamma(t)) = dF(\nabla \|\vec{H}\|(\gamma(0))) \neq 0$ for all $t \in [-a,b]$. But dF is linear, so that $\nabla \|\vec{H}\|(\gamma(t)) \neq 0$ for all $t \in [-a,b]$, thence $\gamma(t) \in U$ for all $t \in [-a,b]$, which contradicts the maximality of $(-a,b)$. So, $\delta(t)$ is defined for all $t \in \mathbb{R}$ and $F \circ i_{\mathcal{F}} \circ \delta$ is a whole line in $\mathbb{R}^{q,n}$. Then, analogous to Lemma 4.1.6 of the first case, $F \circ i_{\mathcal{F}}(\mathcal{F}_p)$ is an affine $m - r$-dimensional subspace of $\mathbb{R}^{q,n}$.

Let us first fix p and prove that $\mathcal{F}_{p'}$ is parallel to $\mathcal{F}_p$ for any $p' \in \mathcal{E}_p$. As $\mathcal{E}_p$ is a 1-dimensional manifold (a curve), we parametrize it by arc length: $\zeta : [0,a] \to i_{\mathcal{E}}(\mathcal{E}_p)$ with $\zeta(0) = p$ and $\zeta(a) = p'$. Then, let $f_p \in \mathcal{F}M_p$ be a vector and $f(t)$ the parallel transport of f_p along ζ with respect to the Levi-Civita connection ∇ of M. From $\frac{d}{dt}\langle \dot\zeta, f(t) \rangle = \langle \nabla_{\dot\zeta} \dot\zeta, f(t) \rangle + \langle \dot\zeta, \nabla_{\dot\zeta} f(t) \rangle = 0$ follows that $\langle \dot\zeta, f(t) \rangle = \langle \dot\zeta(0), f_p \rangle = 0$ and $f(t) \in \mathcal{F}U$ for all $t \in [0,a]$, because $\mathcal{E}U_q$ is one-dimensional and $T_q U = \mathcal{F}U_q \oplus \mathcal{E}U_q$ for any $q \in U$. By writing the second

---

[12]From eq. (4.51) $F \circ \gamma$ is a geodesic of $\mathbb{R}^{q,n}$ and thence a straight line.

fundamental tensor of F like equation (4.49) we get, using eq. (4.48),

$$D_{\dot\zeta} f(t) = \nabla_{\dot\zeta} f(t) + A(\dot\zeta, f(t)) = 0,$$

where D is the Levi-Civita connection of $\mathbb{R}^{q,n}$.

This means that $dF(f(t))$ is also the parallel translation of $dF(f_p)$ in $\mathbb{R}^{q,n}$, which is $dF(f(t)) = dF(f_p)$ for all $t \in [0,a]$. This means that $dF(\mathcal{F}U_p) \subset dF(\mathcal{F}U_{p'})$. Analogously, $dF(\mathcal{F}U_{p'}) \subset dF(\mathcal{F}U_p)$. But we already know that the leaves of $\mathcal{F}U$ are affine subspaces, so that they are equal, up to a translation, to their tangent spaces, and thence are parallel. $\square$

Let us now consider the 1-dimensional leaf of $\mathcal{E}_p$, for some $p \in U$.

Lemma 4.2.7. *Under the hypothesis of Theorem 4.2.1, it holds that the image of $\mathcal{E}_p$ through $F \circ i_{\mathcal{E}_p}$ on $\mathbb{R}^{q,n}$ lies in a plane for every $p \in U$.*

Proof. First of all, $\mathcal{E}_p$ is a 1-dimensional immersed submanifold in $U \subset \mathring{M} \subset M$, under the natural inclusion $i_{\mathcal{E}_p} : \mathcal{E}_p \to U \subset M$, so that $F \circ i_{\mathcal{E}_p} : \mathcal{E}_p \to \mathbb{R}^{q,n}$ is also an immersion and we can consider the second fundamental tensors of them

$$\begin{array}{c} M \xrightarrow{F} \mathbb{R}^{q,n} \\ {\scriptstyle i_{\mathcal{E}_p}} \uparrow \; \nearrow {\scriptstyle F \circ i_{\mathcal{E}_p}} \\ \mathcal{E}_p \end{array},$$

let us drop the index p to make the notation less heavy. So, $A_{F \circ i_\mathcal{E}}$ and $A_{i_\mathcal{E}}$ denote the second fundamental tensors of $F \circ i_\mathcal{E}$ and $i_\mathcal{E}$ respectively. From equation (2.4) we have that

$$A_{F \circ i_\mathcal{E}} = A_F + dF(A_{i_\mathcal{E}}). \tag{4.52}$$

On the other hand, one can write, for vector fields $e_1, e_2 \in \Gamma(T\mathcal{E}_p)$, the second fundamental tensor of the immersion $i_\mathcal{E}$ as

$$A_{i_\mathcal{E}}(e_1, e_2) = \nabla_{e_1} e_2 - \nabla'_{e_1} e_2, \tag{4.53}$$

where ∇' is the Levi-Civita connection of $\mathcal{E}_p$ (with respect to metric induced by the inclusion) and we identify vectors in $T_q\mathcal{E}$ with vectors in T_qM through $di_\mathcal{E}$. But $\nabla_{e_1} e_2 \in \Gamma(\mathcal{E}M)$ by eq. (4.46) and $A_{i_\mathcal{E}}(e_1, e_2) \in \Gamma(T\mathcal{E}^\perp)$, so that equation (4.53) implies that

$$A_{i\varepsilon} = 0, \tag{4.54}$$

so that $\mathcal{E}_p$ is a geodesic of $U \subset \overset{\circ}{M} \subset M$. By Lemma 4.0.28, $\mathcal{E}U_{p'}$ is spanned by $\nabla\|\vec{H}\|/\|\nabla\|\vec{H}\|\|(p')$ for any $p' \in U$. Let $\gamma(s) : (-a,b) \to M$ be the local parametrization by arc lenght of this geodesic in one of the directions $\pm\nabla\|\vec{H}\|/\|\nabla\|\vec{H}\|\|$ with $\gamma(0) = p$. The domain of γ, $(-a,b)$, is open because $\mathcal{E}_p$ is a leaf in the open subset $U \subset M$.

We claim that this curve lies in a plane. First of all

$$\frac{d}{ds}(F \circ \gamma) = dF \circ d\gamma\left(\frac{d}{ds}\right) = \pm dF\left(\frac{\nabla\|\vec{H}\|}{\|\nabla\|\vec{H}\|\|}\right),$$

$$\frac{d^2}{ds^2}(F \circ \gamma) = \left(\nabla_{\frac{\nabla\|\vec{H}\|}{\|\nabla\|\vec{H}\|\|}} dF\right)\left(\frac{\nabla\|\vec{H}\|}{\|\nabla\|\vec{H}\|\|}\right) + dF\left(\nabla_{\frac{\nabla\|\vec{H}\|}{\|\nabla\|\vec{H}\|\|}}\frac{\nabla\|\vec{H}\|}{\|\nabla\|\vec{H}\|\|}\right)$$

$$= A_{ij}\frac{\nabla^i\|\vec{H}\|\nabla^j\|\vec{H}\|}{\|\nabla\|\vec{H}\|\|^2} = \|\vec{H}\|\nu,$$

where we used equation (4.45) and $\nabla_{\dot\gamma}\dot\gamma = 0$. Concerning ν, we get, for any $N \in T_p U^\perp$,

$$\left\langle \nabla_{\pm\frac{\nabla\|\vec{H}\|}{\|\nabla\|\vec{H}\|\|}}\nu, N \right\rangle = 0,$$

because $\nabla^\perp \nu = 0$. For $f \in \Gamma(\mathcal{F}U)$, we get

$$\left\langle \nabla_{\pm\frac{\nabla\|\vec{H}\|}{\|\nabla\|\vec{H}\|\|}}\nu, dF(f) \right\rangle = \pm\frac{\nabla\|\vec{H}\|}{\|\nabla\|\vec{H}\|\|}\langle\nu, dF(f)\rangle - \left\langle\nu, \pm\nabla_{\frac{\nabla\|\vec{H}\|}{\|\nabla\|\vec{H}\|\|}}dF(f)\right\rangle$$

$$= -\left\langle\nu, A\left(\pm\frac{\nabla\|\vec{H}\|}{\|\nabla\|\vec{H}\|\|}, f\right) + dF\left(\nabla_{\pm\frac{\nabla\|\vec{H}\|}{\|\nabla\|\vec{H}\|\|}}f\right)\right\rangle$$

$$= -\frac{1}{\|\vec{H}\|}P\left(\frac{\pm\nabla\|\vec{H}\|}{\|\nabla\|\vec{H}\|\|}, f\right) = 0,$$

where equation (4.45) and the fact that f is in the kernel of P were used.

Finally, identifying vector fields in M and in $R^{q,n}$ through dF,

$$\left\langle \nabla_{\pm\frac{\nabla\|\vec{H}\|}{\|\nabla\|\vec{H}\|\|}}\nu, \frac{\pm\nabla\|\vec{H}\|}{\|\nabla\|\vec{H}\|\|}\right\rangle = \frac{\nabla\|\vec{H}\|}{\|\nabla\|\vec{H}\|\|}\left\langle \nu, \frac{\nabla\|\vec{H}\|}{\|\nabla\|\vec{H}\|\|}\right\rangle - \left\langle \nu, D_{\frac{\nabla\|\vec{H}\|}{\|\nabla\|\vec{H}\|\|}}\frac{\nabla\|\vec{H}\|}{\|\nabla\|\vec{H}\|\|}\right\rangle$$

$$= -\left\langle \nu, A\left(\frac{\nabla\|\vec{H}\|}{\|\nabla\|\vec{H}\|\|}, \frac{\nabla\|\vec{H}\|}{\|\nabla\|\vec{H}\|\|}\right)\right\rangle + \nabla_{\frac{\nabla\|\vec{H}\|}{\|\nabla\|\vec{H}\|\|}}\frac{\nabla\|\vec{H}\|}{\|\nabla\|\vec{H}\|\|}$$

$$= -\frac{1}{\|\vec{H}\|}P\left(\frac{\nabla\|\vec{H}\|}{\|\nabla\|\vec{H}\|\|}, \frac{\nabla\|\vec{H}\|}{\|\nabla\|\vec{H}\|\|}\right) = -\|\vec{H}\|,$$

so that

$$\nabla_{\pm\frac{\nabla\|\vec{H}\|}{\|\nabla\|\vec{H}\|\|}}\nu = -\|\vec{H}\|dF\left(\pm\frac{\nabla\|\vec{H}\|}{\|\nabla\|\vec{H}\|\|}\right). \tag{4.55}$$

Let $\mathcal{H}$ be an (pure real or pure imaginary) antiderivative of $\|\vec{H}\|$ restricted to γ, so that $\dot{\mathcal{H}}(t) = \|\vec{H}\|(\gamma(t))$. Then the family of vectorfields[13] $V_\alpha \in \Gamma((F \circ \gamma)^{-1}(\mathbb{R}^{q,n}))$, for $\alpha \in \mathbb{R}$ if $\|\vec{H}\|^2 > 0$ or $\alpha = i\beta$ with $\beta \in \mathbb{R}$ if $\|\vec{H}\|^2 < 0$, given by

$$V_\alpha := \cos(\mathcal{H}+\alpha)dF\left(\pm\frac{\nabla\|\vec{H}\|}{\|\nabla\|\vec{H}\|\|}\right) - \sin(\mathcal{H}+\alpha)\nu$$

satisfies

$$\frac{d}{dt}(V_\alpha) = -\|\vec{H}\|\sin(\mathcal{H}+\alpha)dF\left(\frac{\pm\nabla\|\vec{H}\|}{\|\nabla\|\vec{H}\|\|}\right) + \cos(\mathcal{H}+\alpha)\|\vec{H}\|\nu$$

$$- \cos(\mathcal{H}+\alpha)\|\vec{H}\|\nu + \sin(\mathcal{H}+\alpha)\|\vec{H}\|dF\left(\pm\frac{\nabla\|\vec{H}\|}{\|\nabla\|\vec{H}\|\|}\right) = 0,$$

this means that any V_α is parallel translated over $F \circ \gamma : \mathbb{R} \to \mathbb{R}^{q,n}$ (thence a constant vector). But V_{α_1} and V_{α_2} are linearly independent for any $\alpha_1 \neq \alpha_2 \in [0, 2\pi)$ (or $i\alpha_1 \neq i\alpha_2 \in [0, 2\pi)$ if $\|\vec{H}\|^2 < 0$) because $\frac{\nabla\|\vec{H}\|}{\|\nabla\|\vec{H}\|\|}$ and $\vec{H}$ are not zero and orthogonal. This implies that $\pm dF\left(\frac{\nabla\|\vec{H}\|}{\|\nabla\|\vec{H}\|\|}\right)$ can be written as a linear combination of two vector fields of the family V_α and lies in the constant plane defined by this family of vector fields and a point of $F \circ \gamma$. By the unicity of

---

[13]Note that although ν can be an imaginary vector field, V_α is real.

the solution (Picard-Lindelöf Theorem) of the O.D.E.

$$\gamma'(t) = \pm dF\left(\frac{\nabla\|\vec{H}\|}{\|\nabla\|\vec{H}\|\|}\right)(t), \qquad \gamma(0) = p,$$

in $\mathbb{R}^{q,n}$ and in the plane defined by the family V_α, it holds that the curve $F \circ \gamma$ lies in this plane, which is orthogonal to $F \circ i_{\mathcal{F}}(\mathcal{F}_p)$ because $\frac{\nabla\|\vec{H}\|}{\|\nabla\|\vec{H}\|\|}$ and $\vec{H}$ are orthogonal to any $f \in \mathcal{F}U$.

$\square$

Remark 4.2.8. Until this point all was done inside an open set $U \subset \mathring{M} \subset M$ such that $\nabla\|\vec{H}\|(p) \neq 0$ for all $p \in U$, we still need to extend these results over the whole manifold.

We want to get a result over the whole manifold M, not only on a connected component U of $\mathring{M}$. For that, we need to take a closer look at the set $M \setminus \mathring{M} = \{p \in M : \nabla\|\vec{H}\|(p) = 0\}$. In $\overline{M}$ the same equations hold as in $\mathring{M}$, so that we only need to look at the open sets of $M \setminus \mathring{M}$.

As $\mathring{\theta} = \theta$ and $\|\mathring{A}\|^2 = \|A\|^2 - \|P * A\|^2$ (eq. (4.44)) in any open sets of $M \setminus \mathring{M}$, Lemma 4.1.3 is proven in this case exactly as Lemma 4.2.3 in any open subset $V \subset M \setminus \mathring{M}$, thence all equations up to eq. (4.27) of the first case also hold in V. Let us consider V maximal, such that $\partial V \subset \partial \mathring{M}$, this implies, for a point $q \in \partial V$, that the tensor P has only one nonzero eigenvalue, and it has to be 1, because on the boundary the equations for $\mathring{M}$ and for V hold (by continuity), but the multiplicity of the eigenvalue 1 in V is constant because $\nabla P = 0$ in this set and P also has only one eigenvector associated with a nonzero eigenvalue in V. Beyond this, in V, $\|\vec{H}\|^2 = \text{tr}(P) = 1$ by eq. (4.12), so that if there is such a nonempty open set V, then $\|\vec{H}\|^2 > 0$ on the whole of M, because $\|\vec{H}\|^2 \neq 0$.

In V, one also gets two orthogonal distributions, $\mathcal{E}'V$ and $\mathcal{F}'V$, which are involutive and totally geodesic; beyond that, the leaves of $\mathcal{F}'V$ are affine spaces, so that $\mathcal{F}'_p \parallel \mathcal{F}'_q$ for any $p, q \in \mathcal{E}'_p \subset V$. In particular, for any $p \in V$, the equations $A_{ij} = P_i^k A_{kj}$ (eq. (4.20)) and $\nabla P = 0$ (from Lemma 4.0.27) hold. We denote the leaves that contain $p \in V$ by $\mathcal{E}'_p$ and $\mathcal{F}'_p$, and their immersions in M by $i_{\mathcal{E}'}$ and $i_{\mathcal{E}'}$ respectively.

We now prove that Lemma 4.2.7 also holds for leaves for the distribution $\mathcal{E}'V$.

Lemma 4.2.9. *Under the hypothesis of Theorem 4.2.1, it holds that the image of $\mathcal{E}'_p$ through $F \circ i_{\mathcal{E}'_p}$ on $\mathbb{R}^{q,n}$ lies in a plane for every $p \in V$.*

Proof. Let $\delta : (a,b) \to V \subset M$ be a maximal, by arc length, parametrization of the 1-dimensional immersed submanifold $\mathcal{E}'_p$ and $\gamma := F \circ \delta$. Then, for any $q' \in i_{\mathcal{E}'}(\mathcal{E}'_p)$, let $\{f_1, \ldots, f_{m-1}\}$ be a basis of $\mathcal{F}M_{q'}$ so that $\{\dot{\gamma}, f_1, \ldots, f_{m-1}\}$ is a basis of $T_{q'}M$ and equation (4.20) implies

$$A(\dot{\gamma}, \dot{\gamma}) = A(P(\dot{\gamma}), \dot{\gamma}) + \sum_i A(P(f_i), f_i) = A(\dot{\gamma}, \dot{\gamma}) + \sum_i A(f_i, f_i) = \vec{H}.$$

then it follows that

$$\frac{d}{dt}(F \circ \gamma) = D_{dF(\dot{\gamma})}dF(\dot{\gamma}) = A(\dot{\gamma}, \dot{\gamma}) + \nabla_{\dot{\gamma}}\dot{\gamma}$$
$$= \vec{H} = \nu,$$

because $\|\vec{H}\|^2 = 1$ in V and

$$\left\langle \frac{d}{dt}\nu, dF(\dot{\gamma}) \right\rangle = dF(\dot{\gamma})\langle \nu, \dot{\gamma} \rangle - \langle \nu, D_{dF(\dot{\gamma})}dF(\dot{\gamma}) \rangle = -\langle \nu, \nu \rangle = -1,$$

but for any $f \in \Gamma(\mathcal{F}'V)$:

$$\left\langle \frac{d}{dt}\nu, dF(f) \right\rangle = dF(\dot{\gamma})\langle \nu, f \rangle - \langle \nu, D_{dF(\dot{\gamma})}dF(f) \rangle$$
$$= -\langle \nu, A(\dot{\gamma}, f) \rangle - \langle \nu, \nabla_{\dot{\gamma}}f \rangle = 0,$$

because (4.20) implies $A(f, X) = 0$ for all $X \in \Gamma(TV)$. Further, for any $N \in \Gamma(TV^\perp)$, it holds that

$$\left\langle \frac{d}{dt}\nu, N \right\rangle = \langle \nabla_{\dot{\gamma}}\nu, N \rangle = 0,$$

due to $\nabla^\perp \nu = 0$, so that

$$\frac{d}{dt}\nu = -\dot{\gamma},$$

by this, one sees that the family of vectorfields $V_\beta \in \Gamma((F \circ \gamma)^{-1}(T\mathbb{R}^{q,n}))$, defined for each $\beta \in \mathbb{R}$ as

$$V_\beta := \cos(t+\beta)dF(\dot{\gamma}) - \sin(t+\beta)\nu$$

is parallel translated in $\mathbb{R}^{q,n}$ and thence $F\circ\gamma$ is contained in the plane defined by two vectorfields of this family and a point of γ (analogous to Lemma 4.2.7). □

Now let us see what the whole M looks like. We saw then, that the tensor P has globally only one non-zero eigenvector and that the eigenspaces of P give globally the distributions

$$\mathcal{E}M := \{V \in M : P(V) = \|\vec{H}\|^2 V\}$$
$$\mathcal{F}M := \{V \in M : P(V) = 0\},$$

which are involutive and whose leaves ($\mathcal{E}_p$ and $\mathcal{F}_p$) are totally geodesic (by different reasons on $\overset{\circ}{M}$ and in $M \setminus \overset{\circ}{M}$) with the inclusion $i_\mathcal{E}$ and $i_\mathcal{F}$. By continuity on the boundary points, all the leaves of $\mathcal{F}M$ are $(m-1)$-dimensional affine spaces of $\mathbb{R}^{q,n}$. Let $\gamma : \mathbb{R} \to M$ be a, by arc length, parametrization of the leaf $\mathcal{E}_p$ for some $p \in M^{14}$, then, for any $f \in \Gamma(\mathcal{F}M)$, $D_{\dot\gamma} f = 0$ (by different reasons on $\overset{\circ}{M}$ and in $M \setminus \overset{\circ}{M}$), and, analogous to the last part of Lemma 4.2.6, all the leaves of $\mathcal{F}M$ that go through points of $\mathcal{E}_p$ are parallel.

Lemma 4.2.10. *Under the hypothesis of Theorem 4.2.1, let $p \in M$ and $\gamma : \mathbb{R} \to M$ be a, by arc length, parametrization of the leaf $\mathcal{E}_p$, then γ lies in a 2-dimensional plane, beyond this the plane is normal to the affine space $F \circ i_\mathcal{F}(\mathcal{F}_q)$, for any $q \in i_\mathcal{E}(\mathcal{E}_p)$.*

Proof. The curve γ lies in a plane. To see this, one defines the family of vectorfields $V_\alpha \in \Gamma((F \circ \gamma)^{-1}(T\mathbb{R}^{q,n}))$ as

$$V_\alpha := \cos(\mathcal{H} + \alpha) dF(\dot\gamma) - \sin(\mathcal{H} + \alpha)\nu,$$

with $\mathcal{H}$ an anti derivative of the restriction of $\|\vec{H}\|$ over γ, then

$$D_{dF(\dot\gamma)} V_\alpha = 0,$$

because of Lemmas 4.2.7 and 4.2.9 using that $\dot\gamma = \pm \frac{\nabla \|\vec{H}\|}{\|\nabla\|\vec{H}\|\|}$ on $\overset{\circ}{M}$ and, as $\|\vec{H}\| = 1 \in \mathbb{R}$ in $M \setminus \overset{\circ}{M}$, $\mathcal{H} = t + k$ (remember that $\|\vec{H}\|^2 > 0$ if $V \neq \emptyset$) in $M \setminus \overset{\circ}{M}$, for some $k \in \mathbb{R}$. Taking two linearly independent vectorfields in this family (Just

---

[14] It is defined for all $\mathbb{R}$ because of the geodesic completeness of M and the totally geodesic property of the leaf

two different α's in $[0, 2\pi)$ because $\dot\gamma$ and ν are always nonzero and orthogonal.) it holds that the curve stays in the plane defined by them. This plane is normal to $F \circ i_{\mathcal{F}}(\mathcal{F}_q)$ for any $q \in \mathcal{E}_p$ because so are ν and $\frac{\nabla\|\vec{H}\|}{\|\nabla\|\vec{H}\|\|}$. □

Let us see what a particular leaf of $\mathcal{E}_p$ looks like.

Lemma 4.2.11. *Under the hypothesis of Theorem 4.2.1, there is a $q \in M$ such that $\mathcal{E}_q$ is a self-shrinker in $\mathbb{R}^{q,n}$, this means that*

$$\vec{H}_{F \circ i_\mathcal{E}}(x) = -(F \circ i_\mathcal{E}(x))^\perp,$$

for every $x \in \mathcal{E}_{q_0}$.

Proof. Because of Remark 4.0.17, we have that $\|F\|^2 \geq \mathcal{E}/2\|F\|_\mathbb{E}^2 > k_1$ outside the euclidean sphere $S^{n-1}(2k_1/\mathcal{E})$ but, by Lemma 4.0.15, the inverse image of $\{X \in \mathbb{R}^{q,n} : \|X\|_\mathbb{E}^2 < 2k_1/\mathcal{E}\}$ through F is contained in a geodesic ball of M. Let $k_1 > \inf_{x \in M} \|F(x)\|^2$, then $\inf_{x \in M}\|F(x)\|^2$ must be assumed by some point inside the geodesic ball, i. e. there is a point $q \in M$, such that $\|F(q)\|^2 = \min_{x \in M}\|F(x)\|^2$. In particular, this implies that

$$0 = \nabla_f \|F\|^2 = \langle \nabla_f F, F\rangle = \langle dF(f), F\rangle,$$

for any $f \in \mathcal{F}M_q$.

Let $\delta : \mathbb{R} \to \mathcal{E}_q$ be a, by arc length, parametrization of the leaf $\mathcal{E}_q$ with $i_\mathcal{E}(\delta(0)) = q$ and write $\gamma := i_\mathcal{E} \circ \delta$. It holds, for any $q' \in i_\mathcal{E}(\mathcal{E}_q)$, that $F \circ i_{\mathcal{F}}(\mathcal{F}_{q'}) \| F \circ i_{\mathcal{F}}(\mathcal{F}_q)$, so that one identifies $f \in \mathcal{F}M_q \cong \mathcal{F}M_{q'} \subset \mathbb{R}^{q,n}$ and calculates

$$\langle F \circ \gamma(t), dF(f)\rangle = \left\langle \int_0^t dF \circ \dot\gamma(s) ds + F \circ \gamma(0), dF(f)\right\rangle$$
$$= \int_0^t \langle dF \circ \dot\gamma(s), dF(f)\rangle ds + \langle F \circ \gamma(0), dF(f)\rangle = 0,$$

because $\dot\gamma(s) \in \mathcal{E}M \perp \mathcal{F}M$, in particular this means that

$$\langle F \circ i_\mathcal{E} \circ \delta(t), dF(f)\rangle = 0 \ \forall t \in \mathbb{R}. \tag{4.56}$$

Then denote $T\mathcal{E}_p^\perp$ the normal bundle of $\mathcal{E}_p$ with respect to the immersion $F \circ i_\mathcal{E}$.

Then eq. (4.56) implies
$$\operatorname{proj}_{T\mathcal{E}_p^\perp}(F\circ\gamma) = (F\circ\gamma)^\perp.$$

Otherwise $A_{ij} = \frac{1}{\|\vec{H}\|}P_{ij}\nu$ (equation (4.45)) in open sets of $\overline{\overset{\circ}{M}}$ and $A_{ij} = P_i^k A_{kj}$ in $M\setminus\overset{\circ}{M}$, so that the only direction that plays a role in the second fundamental tensor is $\dot\gamma$ and

$$\vec{H} = \operatorname{tr}_M A_F = \operatorname{tr}_{\mathcal{E}_p} A_F = \operatorname{tr}_{\mathcal{E}_p} A_{F\circ i_\mathcal{E}} = \vec{H}_{F\circ i_\mathcal{E}},$$

where we used equations (4.26) and (4.24) in the open sets of $M/\overset{\circ}{M}$ and equations (4.52) and (4.54) in $\overset{\circ}{M}$ to get $A_F = A_{F\circ i_\mathcal{E}}$.

This implies that

$$\vec{H}_{F\circ i_\mathcal{E}}(\delta(t)) = \vec{H}(\gamma(t)) = -F^\perp(\delta(t)) = -\operatorname{proj}_{T\mathcal{E}_p^\perp}\gamma(t),$$

so that $i_\mathcal{E}: \mathcal{E}_p \to \mathbb{R}^{q,n}$ is a shrinking self-similar solution of the curve shortening flow. $\square$

Now we prove that $F(M)$ is the product $F(\mathcal{E}_q) \times F(\mathcal{F}_q)$, where $q \in M$ minimizes $\|F\|^2$.

Let $q \in M$ be a minimal point of $\|F\|^2$ and $\{f_1,\ldots,f_{m-r}\}$ be an orthonormal basis of $\mathcal{F}M_q$. We define a function $\mathfrak{h}: \mathcal{E}_q \times \mathbb{R}^{m-r} \to F(M)$, given by

$$\mathfrak{h}(p, X) = F(i_\mathcal{E}(p)) + X^i dF(f_i) \qquad \forall X = (X^1,\ldots, X^{m-r}) \in \mathbb{R}^{m-r}, p\in \mathcal{E}_q.$$

As all the leaves $\mathcal{F}_{q'}$, $q' \in \mathcal{E}_q$, are parallel, the image of $\mathfrak{h}$ is indeed contained in $F(M)$. Let us consider in $\mathbb{R}^{m-r}$ the canonical metric and in $\mathcal{E}_q \times \mathbb{R}^{m-r}$ the product metric, so that $\mathfrak{h}$ is an isometry because F and $i_\mathcal{E}$ are isometries.

$\mathcal{E}_q \times \mathbb{R}^{m-r}$ is geodesically complete (Corollary 6.3.2). We claim that $\mathfrak{h}$ is surjective. To see this, take $(p, X) \in \mathcal{E}_q \times \mathbb{R}^{m-r}$, $y := \mathfrak{h}(p, X) \in F(M)$ and $z \in F(M)$. Let $y' \in M$ and $z' \in M$ be such that $F(y') = y$ and $F(z') = z$. From the fact that M is geodesically complete (hypothesis), there is a vector $Y \in T_{y'}M$ such that $\exp(Y) = z'$ (by the Theorem of Hopf and Rinow, Thr. 6.3.1). Then decompose $Y = Y_1 + Y_2$ with $Y_1 \in T_p\mathcal{E}_q$ and $Y_2 = Y_2^l f_l(p) \in \mathcal{F}M_p$. Now denote

$Y_{20} := (Y_2^1, \ldots, Y_2^{m-r})$, then for the exponential in $\mathcal{E}_q \times \mathbb{R}^{m-r}$ it holds that

$$\mathfrak{h}(\exp(Y_1, Y_{20})) = \exp(d\mathfrak{h}(Y_1, Y_{20})) = \exp(dF \circ di_{\mathcal{E}}(Y_1) + dF(Y_2))$$
$$= F(\exp(di_{\mathcal{E}}(Y_1) + Y_2)) = F(\exp(Y)) = z,$$

where we understand $F(M)$ locally as a manifold (isometric to M and with the same dimension) and thence define the exponential there locally, so that, by the compactness of the domain of the geodesic segment connecting y' and z', the exponential is well defined. This proves that $z \in \mathfrak{h}(\mathcal{E}_q \times \mathbb{R}^{m-r})$.

Then $F(M)$ is the product of an affine space with a shrinking self-similar solution of the mean curvature flow for plane curves.

So we got the product of an affine space with a self-shrinking curve that lies on a plane. Finally, let us take a closer look at each of the factors in this product.

Remark 4.2.12. At the affine space, the induced (from $\mathbb{R}^{q,n}$) inner product has to be positive definite, because we assumed that F is a spacelike immersion.

Note that in $\mathbb{R}^{q,n}$ the way in which the plane lies in the whole space affects the inner product in the plane. From the fact that we considered only spacelike immersions, at least one of the directions in the plane that contains $F \circ i_{\mathcal{E}}(\mathcal{E}_q)$ must be positive definite.

If $\langle \cdot, \cdot \rangle$ restricted to the plane is positive definite one can find a basis made of two orthogonal vectors of length 1, and if one writes the self-shrinking curve in this basis one has just a usual self-shrinker of the curve shortening flow. This is a well studied subject and a classification of such was given by [AL86] and can also be found in [Hal10]. The closed self-shrinkers of the curve shortening flow are called the Abresch & Langer curves, there are also some curves that "do not close" and are dense in some annulus. These curves are not in our classification because they would not satisfy the inverse Lipschitz condition. So that, in our case, the self-shrinking solutions of the mean curvature flow in the plane are just dilatations of the Abresch & Langer curves in $\mathbb{E}^2$.

If $\langle \cdot, \cdot \rangle$ restricted to the plane is degenerate, then the mean curvature vector, which is orthogonal to the tangent direction (which is positive definite), is a null vector, and this is outside the case we are treating ($\|\vec{H}\|^2 \neq 0$).

If $\langle \cdot, \cdot \rangle$ restricted to the plane has one positive and one negative directions, then there could be some different self-shrinkers of the curve shortening flow. But there are none: Ecker showed in [Eck97] the long time existence of spacelike hypersurfaces in the Minkowski space $\mathbb{R}^{1,n}$, but a self-shrinker can only exist for a finite time as eq. (3.1) shows. Then:

Remark 4.2.13. The self-shrinking curve lies in a plane whose induced inner product is positive definite.

$\square$

Remark 4.2.14. In particular, remarks 4.2.12 and 4.2.13 imply that M is contained in the product of an affine space and a plane, both spacelike, so that $\|\vec{H}\|^2 > 0$.

Chapter 5

Summary

In this chapter, we summarize the main results obtained in this work.

First, we found that there are no spacelike self-shrinkers of the mean curvature flow with timelike mean curvature in any of the treated cases, so that Theorems 3.1.3, 4.0.3, 4.0.20, 4.1.1 and Remark 4.2.14 sum up to:

Theorem 5.0.1. *There are no spacelike self-shrinkers $F : M \to (\mathbb{R}^n, \langle \cdot, \cdot \rangle)$ of the MCF that satisfy*

- $F(M)$ *unbounded and F is mainly negative and has bounded geometry or*

- $\|\vec{H}\|^2 < 0$ *and one of the following:*

    1. *M is compact.*

    2. *$F(M)$ is unbounded, M is stochastic complete and $\sup_M \|F\|^2 \leq \infty$.*

    3. *$F(M)$ is unbounded, F is mainly positive, has bounded geometry and the principal normal parallel in the normal bundle.*

Remark 5.0.2. As an immediate consequence there are no spacelike self-shrinking hypersurfaces of the MCF in the Minkowski space $\mathbb{R}^{1,n}$ that are compact or stochastic complete or mainly positive with $\|\vec{H}\|^2 \neq 0$ everywhere.

One expects singularities (at least some of them) to be modeled by self-shrinkers so that this result (which holds for immersions of any codimension) could point to a long time existence of spacelike hypersurfaces with timelike mean curvature vector in a more general context as the one of [Eck97] for hypersurfaces in Minkowski space.

Beyond this, summing up 4.1.1 and 4.2.1, the following classification holds:

Theorem 5.0.3. *Let M be a smooth manifold and $F: M \to \mathbb{R}^{q,n}$ be a mainly positive, spacelike, shrinking self-similar solution of the mean curvature flow with bounded geometry such that $F(M)$ is unbounded. Beyond that, let F satisfy the conditions: $\|\vec{H}\|^2(p) \neq 0$ for all $p \in M$ and the principal normal is parallel in the normal bundle ($\nabla^\perp \nu \equiv 0$). Then one of the two holds:*

$$F(M) = \mathcal{H}_r \times \mathbb{R}^{m-r} \quad \text{or}$$
$$F(M) = \Gamma \times \mathbb{R}^{m-1},$$

where $\mathcal{H}_r$ is an r-dimensional minimal surface of the hyperquadric $\mathcal{H}^{n-1}(r)$ (in addition $\|\vec{H}\|^2 = r > 0$) and Γ is a rescaling of an Abresch & Langer curve in a spacelike plane. By $\mathbb{R}^{m-r}$ we mean an $m-r$ dimensional spacelike affine space in $\mathbb{R}^{q,n}$.

Chapter 6

Appendix

We will now list some of the results we use in the work and do not prove.

6.1 Maximum Principles

Proposition 6.1.1 (Strong Maximum Principle). *Let G be a bounded domain of $\mathbb{R}^n$ and L be an elliptic[1] differential operator, defined for all $u \in C^2(G)$*

$$L[u] := \sum_{i,j=1}^{n} a_{ij} \frac{\partial^2 u}{\partial x^i \partial x^j} + \sum_{i=1}^{n} b_i \frac{\partial u}{\partial x^i}$$

where the coefficients $a_{ij}, b_i : G \to \mathbb{R}$ are continuous functions and suppose that L satisfies the equation

$$L[u] \geq 0.$$

If u has a maximum at an interior point of G, then u is constant in G.

A proof of this can be found in [CH89], page 326.

As a remark it is clear, replacing the function u for $-u$, that the condition $L[u] \leq 0$, is also enough to guarantee that the minimum is at the border or u is constant.

We also use a maximum principle for non-compact Riemannian manifolds, which is equivalent to the stochastic completeness of the manifold.

Definition 6.1.2. A Riemannian Manifold M is said to be *stochastic complete*

---
[1] The quadratic form defined by a_{ij} is positive definite

if, for some (and therefore any) $(x,t) \in M \times (0,+\infty)$, it holds that

$$\int_M \rho(x,y,t)dy = 1,$$

where $\rho(x,y,t)$ is the heat kernel of the Laplacian operator.

Note that this heat kernel is different from the one used in Chapter 4, here it is a fundamental solution of $\frac{d}{dt}u = \triangle u$ which depends on the base manifold and is equal to the heat kernel of Chapter 4 only when the base manifold is flat.

Proposition 6.1.3 (Weak Omori-Yau maximum principle). *Let (M,g) be a smooth, connected, non-compact Riemannian manifold. Then the following are equivalent:*

1. *M is stochastic complete.*

2. *For every $\lambda > 0$, the only non-negative, bounded smooth solution u of $\triangle u = \lambda u$ is $u \equiv 0$.*

3. *For every $u \in C^2(M)$ with $\sup_M u < +\infty$, and for every $\alpha > 0$ set $\Omega_\alpha = \{x \in M : u(x) > \sup_M u - \alpha\}$. Then $\inf_{\Omega_\alpha} \triangle u \leq 0$*

4. *For every $u \in C^2(M)$ with $\sup_M u < +\infty$ there exists a sequence $\{x_n\}$, $n = 1,2,\ldots$, such that, for every n, $u(x_n) \geq \sup_M u - \frac{1}{n}$ and $\triangle u(x_n) \leq \frac{1}{n}$.*

This statement can be found in [PRS03] and has its origins in [Yau75] and [Omo67].

6.2 Foliations

For a more detailed description of foliations, leaves and a proof of the Frobenius Theorem the book [CLN85] can be consulted.

Definition 6.2.1. Let M be a smooth manifold of dimension m. A C^r *foliation* of dimension n of M is a C^r atlas $\mathcal{F}$ on M which is maximal and has the two following properties:

1. If $(U,\varphi) \in \mathcal{F}$ then $\varphi(U) = U_1 \times U_2 \subset \mathbb{R}^n \times \mathbb{R}^{m-n}$ where U_1 and U_2 are open sets in $\mathbb{R}^n$ and $\mathbb{R}^{m-n}$ respectively.

2. If (U,φ) and $(V,\psi) \in \mathcal{F}$ are such that $U \cap V \neq \emptyset$ then the change of coordinates map $\psi \circ \varphi^{-1} : \varphi(U \cap V) \to \psi(U \cap V)$ is of the form

$$\psi \circ \varphi^{-1}(x,y) = (h_1(x,y), h_2(y)).$$

We say that M is *foliated* by $\mathcal{F}$, or that $\mathcal{F}$ is a *foliated structure* of dimension n and class C^r on M.

For $(U,\varphi) \in \mathcal{F}$ like in the last definition the sets of the form $\varphi^{-1}(U_1 \times \{c\})$ with $c \in U_2$ are called *plaques* of $\mathcal{F}$.

A path of plaques of $\mathcal{F}$ is a sequence $\alpha_1, \alpha_2, \ldots, \alpha_k$ of plaques of $\mathcal{F}$ such that $\alpha_j \cap \alpha_{j+1} \neq \emptyset$ for all $j \in \{1, 2\ldots, k-1\}$. Note that M is covered by plaques of $\mathcal{F}$, so that we can define the following equivalence relation:

Definition 6.2.2. Let $p, q \in M$. We say that p is equivalent to q ($p \sim q$) if there exists a path of plaques $\alpha_1, \ldots, \alpha_k$ with $p \in \alpha_1$ and $q \in \alpha_k$. The equivalence classes of $\sim$ are called *leaves* of $\mathcal{F}$.

It is an important remark that every leaf is a differentiable manifold with the atlas induced by the foliation $\mathcal{F}$, which is immersed in M by the inclusion map.

The foliations are related to the following concepts by the Theorem of Frobenius:

Definition 6.2.3. Let M be a smooth manifold. A k-dimensional *distribution* over M is a map $D : M \to TM$ that associates, to each $p \in M$, a k-dimensional subspace of T_pM. A k-dimensional distribution is said to be *differentiable* (of class C^r) if it can be locally spanned by k differentiable (of class C^r) vectorfields. A distribution is said to be *involutive* if, given any two vectorfields $X, Y \in \Gamma(TM)$ with $X(p), Y(p) \in D(p)$ for all $p \in M$, it holds

$$[X,Y](p) \in D(p).$$

Theorem 6.2.4 (of Frobenius). *Let D be a differentiable (of class C^r, $r \geq 1$) k-dimensional distribution on a smooth manifold M. If D is involutive then there exists a C^r foliation $\mathcal{F}$ such that, for any $p \in M$, the leaf $\mathfrak{l}$, that goes through p, satisfies $T_p\mathfrak{l} = D(p)$. Conversely, the tangent bundle of a leaf of a distribution is involutive.*

6.3 Geodesic Completeness

Theorem 6.3.1 (of Hopf and Rinow). *Let M be a smooth Riemannian manifold and $p \in M$ a point. The following statements are equivalent:*

1. *$\exp_p$ is defined for all vectors in $T_p M$.*

2. *Bounded and closed sets are compact.*

3. *M is complete as a metric space.*

4. *M is geodesic complete.*

5. *There is a sequence of compact sets $K_n \in M$, with $K_n \subset K_{n+1} \setminus \partial K_{n+1}$ and $\cup_n K_n = M$, such that, if $q_n \in M$ is a sequence with $q_n \notin K_n$ then $d(p, q_n) \to \infty$ for any $p \in M$.*

Beyond this, any of these implies: For all $q \in M$ there is a geodesic γ connecting p and q such that the length of γ equal to $d(p, q)$.

A proof of this Theorem can be found in [dC92]. As a corollary one gets that the product of geodesic complete manifolds is geodesic complete.

Corollary 6.3.2. *Let M, N be smooth, geodesic complete, Riemannian manifolds, then $M \times N$ (with the product metric) is geodesic complete.*

Proof. Let us use the characterization given by item 3 of the Theorem of Hopf and Rinow. Let $p_n \in M \times N$ be a Cauchy sequence, $p_n = (a_n, b_n)$ with $a_n \in M$ and $b_n \in N$. As the product metric is the sum of the metrics on M and N the projection of a curve in $M \times N$ over M (or N) gives a (piecewise smooth) curve of smaller length. This means that $d(a_n, a_m) \leq d(p_n, p_m)$ and $d(b_n, b_m) \leq d(p_n, p_m)$ so that a_n and b_n are Cauchy sequences on M and N respectively, so that, by the completeness of M and N they converge to $a \in M$ and $b \in N$ respectively and then $p_n \to (a, b)$. □

List of Symbols

A	Second fundamental tensor
D	The Levi-Civita connection of $\mathbb{R}^{q,n}$
$\vec{H}$	The mean curvature vector (field)
$\mathcal{H}^{n-1}(k)$	$\{x \in \mathbb{R}^{q,n} | \langle x, x \rangle = k\}$
P	$P_{ij} := \langle \vec{H}, A_{ij} \rangle$
$P * A$	$P * A_{ij} := P_j^k A_{kj}$
$\mathrm{Pr}_{\mathcal{H}^{n-1}}$	The projection on the tangent bundle of $\mathcal{H}^{n-1}$
Q	$Q_{ij} := \langle A_{ik}, A_j^k \rangle$
R	The Riemannian curvature tensor
S	$S_{ijkl} := \langle A_{ij}, A_{kl} \rangle$
X_+	The projection of $X \in \Gamma(F^*\mathbb{R}^{q,n})$ on the positive directions of $R^{q,n}$
X_-	The projection of $X \in \Gamma(F^*\mathbb{R}^{q,n})$ on the negative directions of $R^{q,n}$
$X_\pm$	X_+ or X_-
$Y^\perp$	The projection of $Y \in \Gamma(F^*\mathbb{R}^{q,n})$ on the normal bundle
$Y^\top$	The projection of $Y \in \Gamma(F^*\mathbb{R}^{q,n})$ on the tangent bundle
θ	$\theta := \frac{1}{2}d\|F\|^2\rangle$
∇	The Levi-Civita connection of M (and the gradient induced by it)
$\triangle$	The (rough) Laplace-Beltrami operator

Bibliography

[AL86] U. Abresch and J. Langer, *The normalized curve shortening flow and homothetic solutions*, J. Differential Geom. **23** (1986), no. 2, 175–196.

[Ale10] Roberta Alessandroni, *Introduction to mean curvature flow*, Actes du Séminaire de Théorie Spectrale et Géométrie. Volume 27. Année 2008–2009, Sémin. Théor. Spectr. Géom., vol. 27, Univ. Grenoble I, Saint, 2010, pp. 1–9.

[Anc06] Henri Anciaux, *Construction of Lagrangian self-similar solutions to the mean curvature flow in $\mathbb{C}^n$*, Geom. Dedicata **120** (2006), 37–48.

[Ang92] Sigurd B. Angenent, *Shrinking doughnuts*, (Gregynog, 1989), Progr. Nonlinear Differential Equations Appl., vol. 7, Birkhäuser Boston, Boston, MA, 1992, pp. 21–38.

[Bak11a] C. Baker, *A partial classification of type I singularities of the mean curvature flow in high codimension*, preprint (2011).

[Bak11b] ———, *The mean curvature flow of submanifolds of high codimension*, preprint (2011).

[BS11] M. Bergner and L. Schäfer, *Time-like surfaces of prescribed anisotropic mean curvature in Minkowski space*, preprint (2011).

[Bra78] Kenneth A. Brakke, *The motion of a surface by its mean curvature*, Mathematical Notes, vol. 20, Princeton University Press, Princeton, N.J., 1978.

[CLN85] César Camacho and Alcides Lins Neto, *Geometric theory of foliations*, Birkhäuser Boston Inc., Boston, MA, 1985. Translated from the Portuguese by Sue E. Goodman.

[CL11] H-D. Cao and H. Li, *A gap theorem for self-shrinkers of the mean curvature flow in arbitrary codimensions*, preprint (2011).

[Cho94] David L. Chopp, *Computation of self-similar solutions for mean curvature flow*, Experiment. Math. **3** (1994), no. 1, 1–15.

[CLN06] Bennett Chow, Peng Lu, and Lei Ni, *Hamilton's Ricci flow*, Graduate Studies in Mathematics, vol. 77, American Mathematical Society, Providence, RI, 2006.

[Coo11] Andrew A. Cooper, *A characterization of the singular time of the mean curvature flow*, Proc. Amer. Math. Soc. **139** (2011), no. 8, 2933–2942.

[CH89] R. Courant and D. Hilbert, *Methods of mathematical physics. Vol. II*, Wiley Classics Library, John Wiley & Sons Inc., New York, 1989. Partial differential equations; Reprint of the 1962 original; A Wiley-Interscience Publication.

[CMI09] T.H. Colding and W.P. Minicozzi II, *Generic mean curvature flow I, generic singularities*, preprint (2009).

[DW09] Q. Ding and Z. Wang, *On the self-shrinking systems in arbitrary codimensional spaces*, preprint (2009).

[dC92] Manfredo Perdigão do Carmo, *Riemannian geometry*, Mathematics: Theory & Applications, Birkhäuser Boston Inc., Boston, MA, 1992. Translated from the second Portuguese edition by Francis Flaherty.

[ES98] Joachim Escher and Gieri Simonett, *The volume preserving mean curvature flow near spheres*, Proc. Amer. Math. Soc. **126** (1998), no. 9, 2789–2796.

[Eck04] Klaus Ecker, *Regularity theory for mean curvature flow*, Progress in Nonlinear Differential Equations and their Applications, 57, Birkhäuser Boston Inc., Boston, MA, 2004.

[Eck97] _____, *Interior estimates and longtime solutions for mean curvature flow of noncompact spacelike hypersurfaces in Minkowski space*, J. Differential Geom. **46** (1997), no. 3, 481–498.

[EH91] Klaus Ecker and Gerhard Huisken, *Parabolic methods for the construction of spacelike slices of prescribed mean curvature in cosmological spacetimes*, Comm. Math. Phys. **135** (1991), no. 3, 595–613.

[EH89] _____, *Mean curvature evolution of entire graphs*, Ann. of Math. (2) **130** (1989), no. 3, 453–471.

[Ger08] Claus Gerhardt, *Curvature flows in semi-Riemannian manifolds*, Surveys in differential geometry. Vol. XII. Geometric flows, Surv. Differ. Geom., vol. 12, Int. Press, Somerville, MA, 2008, pp. 113–165.

[Hal10] H. P. Halldorson, *Self-similar solutions to the curve shortening flow*, preprint (2010).

[Hui84] Gerhard Huisken, *Flow by mean curvature of convex surfaces into spheres*, J. Differential Geom. **20** (1984), no. 1, 237–266.

[Hui90] _____, *Asymptotic behavior for singularities of the mean curvature flow*, J. Differential Geom. **31** (1990), no. 1, 285–299.

[Hui93] _____, *Local and global behaviour of hypersurfaces moving by mean curvature*, Differential geometry: partial differential equations on manifolds (Los Angeles, CA, 1990), Proc. Sympos. Pure Math., vol. 54, Amer. Math. Soc., Providence, RI, 1993, pp. 175–191.

[HS09] Gerhard Huisken and Carlo Sinestrari, *Mean curvature flow with surgeries of two-convex hypersurfaces*, Invent. Math. **175** (2009), no. 1, 137–221.

[Hua11] Rongli Huang, *Lagrangian mean curvature flow in pseudo-Euclidean space*, Chin. Ann. Math. Ser. B **32** (2011), no. 2, 187–200.

[Ilm97a] Tom Ilmanen, *Singularities of mean curvature flow of surfaces*, preprint (1997).

[Ilm97b] _____, *Lectures on mean curvature flow and related equations*, Swiss Federal Institute of Technology, Department of Mathematics (1997).

[JLT10] Dominic Joyce, Yng-Ing Lee, and Mao-Pei Tsui, *Self-similar solutions and translating solitons for Lagrangian mean curvature flow*, J. Differential Geom. **84** (2010), no. 1, 127–161.

[LW10] Yng-Ing Lee and Mu-Tao Wang, *Hamiltonian stationary cones and self-similar solutions in higher dimension*, Trans. Amer. Math. Soc. **362** (2010), no. 3, 1491–1503.

[LS09] Guanghan Li and Isabel M. C. Salavessa, *Mean curvature flow and Bernstein-Calabi results for spacelike graphs*, Differential geometry, World Sci. Publ., Hackensack, NJ, 2009, pp. 164–174.

[LW12] Haizhong Li and Yong Wei, *Classification and rigidity of self-shrinkers in the Mean curvature flow*, preprint (2012).

[Man10] Carlo Mantegazza, *Lecture Notes on Mean Curvature Flow*, Progress in mathematics, 290, Birkhäuser Verlag AG, Basel, Ch, 2010.

[Mul56] W. W. Mullins, *Two-dimensional motion of idealized grain boundaries*, J. Appl. Phys. **27** (1956), 900–904.

[Nev11] A. Neves, *Recent Progress on Singularities of Lagrangian Mean Curvature Flow*, preprint (2011).

[Omo67] Hideki Omori, *Isometric immersions of Riemannian manifolds*, J. Math. Soc. Japan **19** (1967), 205–214.

[O'N83] Barrett O'Neill, *Semi-Riemannian geometry*, Pure and Applied Mathematics, vol. 103, Academic Press Inc. [Harcourt Brace Jovanovich Publishers], New York, 1983. With applications to relativity.

[PRS03] Stefano Pigola, Marco Rigoli, and Alberto G. Setti, *A remark on the maximum principle and stochastic completeness*, Proc. Amer. Math. Soc. **131** (2003), no. 4, 1283–1288 (electronic).

[RS10] Manuel Ritoré and Carlo Sinestrari, *Mean curvature flow and isoperimetric inequalities*, Advanced courses in mathematics CRM Barcelona, Birkhäuser Verlag AG, Basel, Ch, 2010.

[Smo05] Knut Smoczyk, *Self-shrinkers of the mean curvature flow in arbitrary codimension*, Int. Math. Res. Not. **48** (2005), 2983–3004.

[SW11] Knut Smoczyk and Mu-Tao Wang, *Generalized Lagrangian mean curvature flows in symplectic manifolds*, Asian J. Math. **15** (2011), no. 1, 129–140.

[Smo11a] Knut Smoczyk, *Mean curvature flow in higher codimension - introduction and survey*, preprint (2011).

[Smo11b] _____, *On algebraic selfsimilar solutions of the mean curvature flow*, Analysis (Munich) **31** (2011), no. 1, 91–102.

[Top98] Peter Topping, *Mean curvature flow and geometric inequalities*, J. Reine Angew. Math. **503** (1998), 47–61.

[Yau75] Shing Tung Yau, *Harmonic functions on complete Riemannian manifolds*, Comm. Pure Appl. Math. **28** (1975), 201–228.

[Wan09] Lu Wang, *A Bernstein type theorem for self-similar shrinkers*, preprint (2009).

[Wan08a] Mu-Tao Wang, *Lectures on mean curvature flows in higher codimensions*, Handbook of geometric analysis. No. 1, Adv. Lect. Math. (ALM), vol. 7, Int. Press, Somerville, MA, 2008, pp. 525–543.

[Wan08b] _____, *Some recent developments in Lagrangian mean curvature flows*, Surveys in differential geometry. Vol. XII. Geometric flows, Surv. Differ. Geom., vol. 12, Int. Press, Somerville, MA, 2008, pp. 333–347. MR2488942 (2009m:53181)

[Whi94] Brian White, *Partial regularity of mean-convex hypersurfaces flowing by mean curvature*, Internat. Math. Res. Notices **4** (1994), 186 ff., approx. 8 pp. (electronic).

[Whi02] _____, *Evolution of curves and surfaces by mean curvature*, (Beijing, 2002), Higher Ed. Press, Beijing, 2002, pp. 525–538.

[Zhu02] Xi-Ping Zhu, *Lectures on mean curvature flows*, AMS/IP Studies in Advanced Mathematics, vol. 32, American Mathematical Society, Providence, RI, 2002.

Index

Bounded geometry, 75

Distribution, 122

Foliation, 121

Grows polynomially, 75

Heat kernel, 79
Homothety of the MCF, 35
Hyperquadric, 18
hyperquadric, 36

Inner product, 13
Isometric immersion, 22

Laplace Beltrami operator, 24

Mainly negative, 73
Mainly positive, 73
Mean curvature vector, 22
Minimal immersion, 37
Minkowski space, 14

Principal normal, 45

Riemannian curvature tensor, 26

Second fundamental tensor, 22
Self-shrinker, 52
Self-similar shrinking solution, 52
Semi-Riemannian manifold, 20
Semi-Riemannian metric, 20
Solution of the mean curvature flow, 35

Spacelike, 16
Stochastic complete, 120

Unbounded, 70

i want morebooks!

Buy your books fast and straightforward online - at one of world's fastest growing online book stores! Environmentally sound due to Print-on-Demand technologies.

Buy your books online at
www.get-morebooks.com

Kaufen Sie Ihre Bücher schnell und unkompliziert online – auf einer der am schnellsten wachsenden Buchhandelsplattformen weltweit! Dank Print-On-Demand umwelt- und ressourcenschonend produziert.

Bücher schneller online kaufen
www.morebooks.de

VDM Verlagsservicegesellschaft mbH
Heinrich-Böcking-Str. 6-8   Telefon: +49 681 3720 174   info@vdm-vsg.de
D - 66121 Saarbrücken   Telefax: +49 681 3720 1749   www.vdm-vsg.de

Printed by Books on Demand GmbH, Norderstedt / Germany